国外优秀物理著作
原　版　系　列

先进的托卡马克稳定性理论

Advanced Tokamak Stability Theory

（英文）

[美] 郑林锦（Linjin Zheng）著

哈尔滨工业大学出版社
HARBIN INSTITUTE OF TECHNOLOGY PRESS

黑版贸审字 08-2020-160 号

图书在版编目(CIP)数据

先进的托卡马克稳定性理论=Advanced Tokamak Stability Theory:英文/(美)郑林锦(Linjin Zheng)著.—哈尔滨:哈尔滨工业大学出版社,2021.6

ISBN 978-7-5603-9433-6

Ⅰ.①先… Ⅱ.①郑… Ⅲ.①托卡马克装置-稳定性-英文 Ⅳ.①TL631.2

中国版本图书馆 CIP 数据核字(2021)第 089104 号

策划编辑 刘培杰
责任编辑 刘家琳 李 烨
封面设计 孙茵艾
出版发行 哈尔滨工业大学出版社
社 址 哈尔滨市南岗区复华四道街 10 号 邮编 150006
传 真 0451-86414749
网 址 http://hitpress. hit. edu. cn
印 刷 哈尔滨市工大节能印刷厂
开 本 720 mm×1 000 mm 1/16 印张 10.75 字数 197 千字
版 次 2021 年 6 月第 1 版 2021 年 6 月第 1 次印刷
书 号 ISBN 978-7-5603-9433-6
定 价 88.00 元

Contents

Preface

One of the most famous quotes from Albert Einstein is 'The most incomprehensible thing about the world is that it is comprehensible'. Indeed, nature is governed by laws. Our scientific endeavors become meaningful because there are natural laws. Remarkably, even the scientific discoveries of mankind could be considered to follow a law: 'In the sweat of thy face shalt thou eat bread' (Genesis 3:19). The more 'bread' a discovery yields for mankind, the more 'sweat' it will cost mankind to discover it. The most serious challenges humans are facing are food shortages and diminishing energy resources. If natural photosynthesis could be reproduced, the issue of food shortages would be addressed. If controlled nuclear fusion could be achieved with a net energy yield, the energy resource problem would be solved. Unfortunately, the difficulties of achieving these goals are disproportionately large compared to discoveries which are less critically important or which can be harmful to the natural environment.

The most difficult aspect of tokamak plasma theory for controlled nuclear fusion lies in the fact that the simplicity and beauty underlying the philosophy of physics are often lost. Due to the complexity of many charged-particle problems with long correlation lengths and the complications related to toroidal geometry, tokamak physics is intrinsically difficult. Thanks to the decades of theoretical and experimental efforts in this field, tokamak physics has become more comprehensible than ever before. This book gives an overview of these developments, with an emphasis on stability theories in toroidal geometry.

The intention of this book is to introduce advanced tokamak stability theory. The readers are expected to have a basic background in plasma physics. We start with the derivation of the Grad–Shafranov equation and the construction of various toroidal flux coordinates. An analytical tokamak equilibrium theory is presented to demonstrate the Shafranov shift and how the toroidal hoop force can be balanced by the application of a vertical magnetic field in tokamaks. In addressing ideal magnetohydrodynamics (MHD) stability theory, this book starts with the advanced but most fundamental topic: the structure of ideal MHD equations and the MHD mode spectrum. This is followed by the toroidal theories for interchange modes, ballooning modes, peeling and free boundary ballooning modes, toroidal Alfvén eigenmodes (TAEs) and the kinetically driven modes, such as kinetic ballooning modes and energetic particle modes. For resistive MHD stability theory, we first describe the resistive MHD singular layer theory of Glasser, Green and Johnson, followed by the resistive ballooning and tearing mode theories. The gyrokinetic theory is then introduced. Both the electrostatic and electromagnetic modes are discussed under the gyrokinetic framework. First, the fixed and free boundary ion temperature gradient mode theories are presented. Then, the effects of the finite Larmor radius on the toroidal interchange modes are investigated, followed by the kinetic ballooning theory and a discussion of the kinetic TAE theories.

In addition to these advanced theories, this book also discusses the intuitive physics pictures for various experimentally observed phenomena. These basic pictures are particularly important for understanding tokamak discharges. First, a schematic classification is described for various tokamak confinement modes, such as the low and high confinement modes (L- and H-modes) and the improved energy confinement mode (I-mode). Then, the physical interpretations are discussed for various core and edge stability/transport phenomena, such as the transport barrier, nonlocal transport, edge localized modes, blob transport and edge harmonic oscillations.

With the completion of ITER in sight and the rapid developments in tokamak diagnostic techniques and discharge control methods, great challenges still remain for plasma physics theoreticians. Many critical issues remain open in this field. I hope this book proves helpful in promoting further efforts to address these issues.

Dr Linjin Zheng
Institute for Fusion Studies
The University of Texas at Austin
Texas, USA
31 December 2014

Acknowledgments

This book was written where the author currently works, at the Institute for Fusion Studies, The University of Texas at Austin, with support from the US Department of Energy, Office of Fusion Energy Science: Grant No. DE-FG02-04ER-54742. Research results in this book have benefited from collaborations with colleagues at the institute, including Drs M T Kotschenreuther, H L Berk, F L Waelbroeck, R D Hazeltine, B Breizman, W Horton, *et al*, and J W Van Dam, who is now at the US Department of Energy. The author would also like to acknowledge collaborations with many other colleagues in the field of plasma physics.

Author biography

Linjin Zheng

Dr Linjin Zheng is a theoretical physicist in the field of controlled thermonuclear fusion plasmas. He received his MSc degree from The University of Science and Technology of China and his PhD from the Institute of Physics, Beijing, Chinese Academy of Sciences. He currently works at Institute for Fusion Studies, The University of Texas at Austin. He has published more than a hundred scientific papers, for example in *Physical Review Letters*, *Physics Letters*, *Nuclear Fusion*, *Physics of Plasmas* and major conferences. His research covers tokamak equilibrium and stability theories in ideal/resistive magnetohydrodynamics, two fluids and kinetics. His major contributions with colleagues include the reformulation of gyrokinetic theory, the development of the theoretical interpretation for the so-called edge localized modes, the invention of the free boundary ballooning representation, the discoveries of the second toroidal Alfvén eigenmodes and current interchange tearing modes, etc. He has also developed the AEGIS and AEGIS-K codes.

Linjin Zheng

Chapter 1

Tokamak MHD equilibrium

In this chapter, we will describe the tokamak magnetohydrodynamics (MHD) equilibrium theory. The basic set of ideal MHD equations are derived from the single-fluid and Maxwell's equations. They are given as follows [1][1]

$$\rho \frac{\mathrm{d}\mathbf{v}}{\mathrm{d}t} = -\nabla P + \mathbf{J} \times \mathbf{B}, \tag{1.1}$$

$$\mathbf{E} = -\mathbf{v} \times \mathbf{B}, \tag{1.2}$$

$$\frac{\partial P}{\partial t} = -\mathbf{v} \cdot \nabla P - \Gamma P \nabla \cdot \mathbf{v}, \tag{1.3}$$

$$\frac{\partial \rho_m}{\partial t} = -\mathbf{v} \cdot \nabla \rho_m - \rho_m \nabla \cdot \mathbf{v}, \tag{1.4}$$

$$\mu_0 \mathbf{J} = \nabla \times \mathbf{B}, \tag{1.5}$$

$$\frac{\partial \mathbf{B}}{\partial t} = \nabla \times \mathbf{E}, \tag{1.6}$$

where ρ_m is mass density, $\mathbf{v}$ is fluid velocity, P is plasma pressure, Γ represents the ratio of specific heats, $\mathbf{E}$ and $\mathbf{B}$ represent the electric and magnetic fields, respectively, $\mathbf{J}$ is current density, μ_0 is vacuum permeability, and bold font denotes vectors.

[1] Some results in this book refer to the author's personal notes and private communications, which are not listed in detail in the reference lists.

doi:10.1088/978-1-6270-5423-2ch1

The MHD equilibrium equations can be obtained from this set of MHD equations as follows:

$$\mathbf{J} \times \mathbf{B} = \nabla P, \tag{1.7}$$

$$\mu_0 \mathbf{J} = \nabla \times \mathbf{B}, \tag{1.8}$$

$$\nabla \cdot \mathbf{B} = 0. \tag{1.9}$$

One can obtain general features of the equilibrium current from the basic set of equilibrium equations (1.7)–(1.9). Due to the existence of the plasma pressure gradient, a diamagnetic current is present in tokamak systems. From the force balance equation (1.7) one can find that the total equilibrium current can be expressed as

$$\mathbf{J} = \sigma \mathbf{B} + \frac{\mathbf{B} \times \nabla P}{B^2}, \tag{1.10}$$

where σ represents the parallel current and the second term on the right is the diamagnetic current. The diamagnetic current itself is not divergence-free and is always accompanied by a return current in the parallel direction. The return current is referred to as the Pfirsch–Schlüter current. One can therefore determine the Pfirsch–Schlüter current from (1.10), by requiring the total current to be divergence-free, i.e. $\nabla \cdot \mathbf{J} = 0$,

$$\sigma = -\int_0^l \nabla \times \frac{\mathbf{B}}{B^2} \cdot \nabla P \frac{\mathrm{d}l}{B} + \sigma_0, \tag{1.11}$$

where l is the arc length along the magnetic field line and σ_0 is the integration constant. Physically, σ_0 specifies the parallel Ohmic current and therefore can be determined by Ohm's law in the parallel direction.

In the following sections, we will first derive the fundamental tokamak equilibrium equation, the Grad–Shafranov equation. Then, we will describe the construction of various toroidal flux coordinates. Finally, we will discuss the analytical solution for a tokamak equilibrium with a large aspect ratio and a circular cross section, to investigate how the hoop force and vertical magnetic field can affect the equilibrium.

1.1 The Grad–Shafranov equation

One of the most important features of tokamaks is toroidal symmetry. This feature leads to the existence of magnetic surfaces in the tokamak equilibrium. In this section, we outline the derivation of the Grad–Shafranov equation [2, 3].

As shown in figure 1.1, the cylindrical coordinate system (R, ϕ, Z) is introduced in the rectangular coordinate system (X, Y, Z), where Z denotes the axisymmetry axis of the plasma torus, R is the major radius from the axis, and ϕ is the toroidal axisymmetric angle. In figure 1.1, the quasi-toroidal coordinate system (r, θ, ϕ) is also shown, in which r is the minor radius and θ represents the poloidal angle. To maintain both the (R, ϕ, Z) and (r, θ, ϕ) coordinate systems to be right handed, the axisymmetry axis Z is chosen as pointing downward for theoretical convenience.

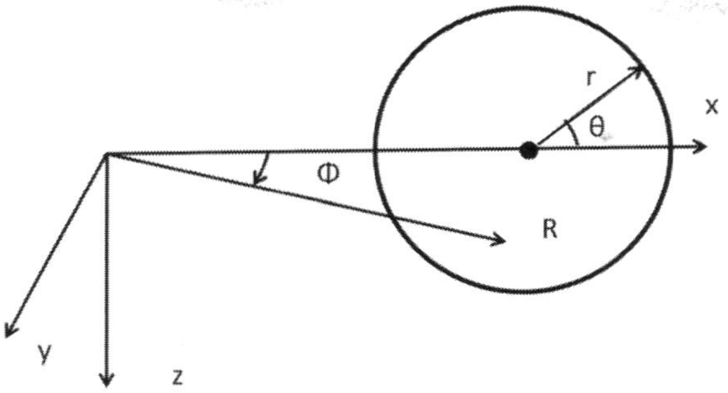

Figure 1.1. Coordinates.

We introduce the vector potential $\mathbf{A}$ to represent the magnetic field $\mathbf{B} = \nabla \times \mathbf{A}$. Due to toroidal symmetry the toroidal angle ϕ is an ignorable coordinate. Using the curl expression in cylindrical coordinates and noting that $\partial A_R/\partial\phi = \partial A_Z/\partial\phi = 0$, one can prove that the two components of the vector potential $\mathbf{A}$ on the ϕ = const. plane (A_R, A_Z) can be expressed through a single toroidal component: $\mathbf{A}_\phi$. Without losing generality, one can write $\mathbf{A}_\phi = -\chi\nabla\phi$. Therefore, the total equilibrium magnetic field can be expressed, by adding the toroidal component B_ϕ, as

$$\mathbf{B} = \nabla \times \mathbf{A}_\phi + RB_\phi\nabla\phi = \nabla\phi \times \nabla\chi + g\nabla\phi, \tag{1.12}$$

where $g = RB_\phi$. From (1.12) one can prove that $\mathbf{B} \cdot \nabla\chi = 0$ and therefore χ = const. marks the magnetic surfaces. Equation (1.12) can be used to show that $2\pi\chi$ represents poloidal magnetic flux. One can also define the toroidal flux $2\pi\psi(\chi)$. The safety factor is then defined as $q = \mathrm{d}\psi/\mathrm{d}\chi$, which characterizes the field line winding on a magnetic surface.

Using Ampere's law in (1.8) one can express equilibrium current density as follows, using (1.12),

$$\mu_0\mathbf{J} = \nabla g \times \nabla\phi + R^2\nabla \cdot \left(\frac{\nabla\chi}{R^2}\right)\nabla\phi. \tag{1.13}$$

Here, we have noted that

$$\nabla\phi \cdot \nabla \times (\nabla\phi \times \nabla\chi) = \nabla \cdot \left[(\nabla\phi \times \nabla\chi) \times \nabla\phi\right] = \nabla \cdot \left(\nabla\chi/R^2\right),$$

$$\nabla\theta \cdot \nabla \times (\nabla\phi \times \nabla\chi) = \nabla\chi \cdot \nabla \times (\nabla\phi \times \nabla\chi) = 0.$$

From the force balance equation (1.7) one can find that $\mathbf{B} \cdot \nabla P = 0$. Therefore, one can conclude that the plasma pressure is a surface function, i.e. $P = P(\chi)$. From (1.7) one can further obtain that $\mathbf{J} \cdot \nabla P = P'_\chi \mathbf{J} \cdot \nabla\chi = 0$, where the prime is used to denote the derivative with respect to the corresponding flux coordinate. Therefore, one can further determine that g is a surface function also by projecting (1.13) on $\nabla\chi$.

Inserting (1.12) and (1.13) into the force balance equation (1.7) and projecting the resulting equation onto the $\nabla\chi$-direction, one obtains the so-called Grad–Shafranov equation

$$R^2\nabla \cdot \left(\frac{\nabla\chi}{R^2}\right) = -\mu_0 R^2 P'_\chi - gg'_\chi. \tag{1.14}$$

Incidentally, using (1.12), (1.13) and (1.14) one can express the current density parallel to the magnetic field as

$$J_\| = -\frac{\mu_0 g P'_\chi}{B} - g'_\chi B. \tag{1.15}$$

The Grad–Shafranov equation is a nonlinear equation. By specifying two free functions $P(\chi)$ and $g(\chi)$ one can find the solution for χ. This generally requires a numerical solution. Since this is a two-dimensional problem, one needs to introduce a poloidal angle coordinate θ_{eq} around the magnetic axis of the plasma torus in addition to the radial flux coordinate χ. The solution is usually given on the $\phi = 0$ plane. There are two types of expressions: one is obtained by giving $X(\chi, \theta_{eq})$ and $Z(\chi, \theta_{eq})$ in the (χ, θ_{eq}) grids; and the other is expressed inversely by giving $\chi(X, Z)$ and $\theta_{eq}(X, Z)$ in the (X, Z) grids.

1.2 Toroidal flux coordinates

In the previous section, we discussed the Grad–Shafranov equation and its solution. The solution is usually given in cylindrical coordinates, directly or inversely. Note that neither the coordinates (X, Z, ϕ) nor $(\chi, \theta_{eq}, \phi)$ are the so-called magnetic flux coordinates. In most equilibrium codes, θ_{eq} is just an equal-arc-length poloidal coordinate. The magnetic flux coordinates, often used in theoretical analyses, are characterized by a straight magnetic field line in the covariant representation of the coordinate system and $\nabla \cdot \mathbf{B} = 0$ is satisfied automatically. In this section, we discuss various flux coordinate systems.

One of the flux coordinate systems is the so-called PEST coordinate system [4] $(\chi, \theta_{PEST}, \phi)$, where θ_{PEST} is the generalized poloidal coordinate, such that the equilibrium magnetic field can be represented as

$$\mathbf{B} = \chi'(\psi_{PEST})\big(\nabla\phi \times \nabla\psi_{PEST} + q\nabla\psi_{PEST} \times \nabla\theta_{PEST}\big). \tag{1.16}$$

The key feature of the PEST coordinate system is that the toroidal axisymmetric angle ϕ is kept as the toroidal angle.

By equating (1.16) and (1.12) one can find that the Jacobian of the PEST coordinates should be

$$\mathcal{J}_{PEST} \equiv \frac{1}{\nabla\psi_{PEST} \times \nabla\theta_{PEST} \cdot \nabla\phi} = \frac{q\chi' X^2}{g}. \tag{1.17}$$

In the PEST coordinate system, the flux coordinate is chosen as

$$\psi_{\mathrm{PEST}} = \frac{2\pi X_0}{c_{\mathrm{PEST}}}\int_0^{\chi} \mathrm{d}\chi \frac{q}{g}, \qquad c_{\mathrm{PEST}} = \frac{X_0}{2\pi}\int_v \mathrm{d}\tau \frac{1}{X^2},$$

where X_0 is the major radius at the magnetic axis and $\int_v \mathrm{d}\tau$ denotes volume integration over the entire plasma domain. The PEST poloidal angle θ_{PEST} can be related to the physical poloidal angle θ_{eq} as follows. Note that $\mathcal{J}_{\mathrm{eq}} = 1/\nabla\chi \times \nabla\theta_{\mathrm{eq}} \cdot \nabla\phi = \mathcal{J}_{\mathrm{PEST}}(\partial\theta_{\mathrm{PEST}}/\partial\theta_{\mathrm{eq}})(\mathrm{d}\psi_{\mathrm{PEST}}/\mathrm{d}\chi)$ and $\mathcal{J}_{\mathrm{eq}}$ can be computed from the equilibrium solution. Using (1.17), one can determine the poloidal angle in PEST coordinates

$$\theta_{\mathrm{PEST}} = \frac{1}{q}\int_0^{\theta_{\mathrm{eq}}} \mathrm{d}\theta_{\mathrm{eq}} \frac{g\mathcal{J}_{\mathrm{eq}}}{X^2},$$

where the integration is along the path of constants χ and ϕ.

Next, we discuss the construction of general flux coordinates. The covariant type of representation, as in (1.16), is not unique. Under the following coordinate transformation

$$\zeta = \phi + \nu(\psi, \theta), \qquad \theta = \theta_{\mathrm{PEST}} + \nu(\psi, \theta)/q, \qquad \text{and} \qquad \psi = \psi(\chi), \tag{1.18}$$

the covariant representation in (1.16) is preserved as

$$\mathbf{B} = \chi'(\psi)(\nabla\zeta \times \nabla\psi + q\nabla\psi \times \nabla\theta). \tag{1.19}$$

Here, θ and ζ are referred to as the generalized poloidal and toroidal angles, respectively. It is apparent that PEST is just one of the possible flux coordinate systems.

The general coordinate system can be constructed from the solution of the Grad–Shafranov equation in cylindrical coordinates as follows. By equating (1.19) and (1.12) in the $\nabla\phi$-direction one can find that

$$\left.\frac{\partial\nu}{\partial\theta_{\mathrm{eq}}}\right|_{\psi,\phi} \frac{1}{\mathcal{J}_{\mathrm{eq}}} + g\frac{1}{X^2} = \chi' q \frac{1}{\mathcal{J}}, \tag{1.20}$$

where $\mathcal{J} = 1/\nabla\psi \times \nabla\theta \cdot \nabla\zeta$. Using the $\mathcal{J}_{\mathrm{eq}}$ and $\mathcal{J}$ definitions, one can prove that

$$\left.\frac{\partial\theta}{\partial\theta_{\mathrm{eq}}}\right|_{\psi,\phi} = \frac{\mathcal{J}_{\mathrm{eq}}}{\chi'\mathcal{J}}. \tag{1.21}$$

One can solve (1.20), yielding

$$\nu(\psi, \theta) = \int_0^{\theta_{\mathrm{eq}}} \mathrm{d}\theta_{\mathrm{eq}}\mathcal{J}_{\mathrm{eq}}\left(q\frac{1}{\mathcal{J}} - g\frac{1}{X^2}\right) = q\theta - \int_0^{\theta_{\mathrm{eq}}} \mathrm{d}\theta_{\mathrm{eq}}\frac{g\mathcal{J}_{\mathrm{eq}}}{X^2}, \tag{1.22}$$

where (1.21) has been used and the integration is on the path with constant ψ and ϕ.

Equations (1.18)–(1.22) can be used to construct various types of flux coordinate systems. There are two classes of these: one is obtained by specifying the Jacobian (e.g. Hamada coordinates [5] and Boozer coordinates [6]) and the other by directly choosing the generalized poloidal angle (e.g. the equal-arc-length coordinate). In the Hamada coordinates the volume inside a magnetic surface is used to label the magnetic surfaces, i.e. $\psi = V$, and the Jacobian $\mathcal{J}_{\rm H} = 1/\nabla V \cdot \nabla\theta_{\rm H} \times \nabla\zeta_{\rm H}$ is set to be unity. With the Jacobian specified, (1.21) can be used to solve for $\theta_{\rm H}$ at given (V,ϕ). With ν determined by (1.22) the definition (1.18) can be used to specify $\zeta_{\rm H}$. In the Boozer coordinates the Jacobian is chosen to be $\mathcal{J}_{\rm B} = V'\langle B^2\rangle_{\rm s}/(4\pi^2 B^2)$, where $\langle \cdot \rangle_{\rm s}$ represents the surface average. The procedure for specifying the Boozer poloidal and toroidal coordinates $\theta_{\rm B}$ and $\zeta_{\rm B}$ is similar to that for Hamada coordinates. In the equal-arc-length coordinate system the poloidal angle is directly specified as the equal-arc-length coordinate $\theta_{\rm e}$. In this case, its Jacobian $\mathcal{J}_{\rm e}$ can be computed through (1.21) and ν can be determined by (1.22). Finally, the definition (1.18) can be used to specify the generalized toroidal coordinate $\zeta_{\rm e}$.

We can also express the current density vector in the covariant representation using the generalized flux coordinates. Using Ampere's law in (1.8) for determining $\mathbf{J} \cdot \nabla\theta$ and the force balance equation (1.7) for $\mathbf{J} \cdot \nabla\zeta$, one can express the equilibrium current density in the covariant representation

$$\mathbf{J} = -\frac{1}{\mu_0} g'_\psi \nabla\zeta \times \nabla\psi - \left(\frac{q}{\mu_0} g'_\psi + \frac{P'_\psi}{\chi'_\psi}\mathcal{J}\right)\nabla\psi \times \nabla\theta. \tag{1.23}$$

This general coordinate expression for $\mathbf{J}$ can alternatively be obtained from (1.13) and the Grad–Shafranov equation (1.14) through coordinate transformation. Equation (1.23) is significantly simplified in the Hamada coordinates. As $\mathcal{J} = 1$ in the Hamada coordinates, (1.23) can be reduced to

$$\mathbf{J} = J'_V \nabla\zeta \times \nabla V + I'_V \nabla V \times \nabla\theta, \tag{1.24}$$

where $I(V)$ and $J(V)$ are the toroidal and poloidal current fluxes, $I' = -g'_V/\mu_0$ and $J' = -qg'_V/\mu_0 - P'_V/\chi'_V$. The force balance equation (1.7) can be simply expressed as

$$\mu_0 P'_V = J'_V \psi'_V - I'_V \chi'_V. \tag{1.25}$$

1.3 The tokamak force balance: the hoop force

Since the tokamak concept was first conceived [3], it has been recognized that the hoop force is present in the tokamak force balance. From figure 1.2 one can see that, when the plasma column is bent into a torus, there is a net force pointing outward from the axisymmetric axis. The net force results from both the plasma and magnetic pressures. For example, as they push along the toroidal direction, the plasma and magnetic pressures give rise to a net outward force. Also, as the poloidal magnetic field is stronger on the inner (or high field) side than on the outer (or low field) side, it also pushes the plasma torus outward. To balance this outward hoop force, a vertical magnetic field is usually introduced in the tokamak system. Because

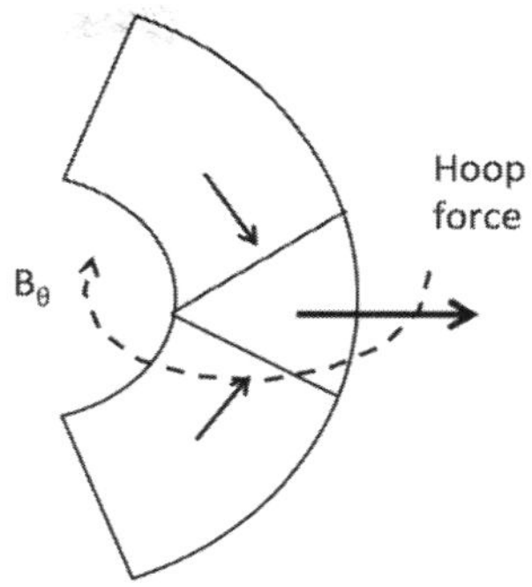

Figure 1.2. Hoop force.

toroidal current exists in the tokamak, the vertical magnetic field can interact with the toroidal current to generate an inward magnetic force.

In this section, we use the analytic tokamak equilibrium solution with a large aspect ratio and a circular cross section to demonstrate the tokamak force balance. The MHD equilibrium equations, (1.7)–(1.9), have been solved for this case in [7]. We extend this solution by including vertical magnetic field effects to show how the vertical field can be used in the tokamak to counterbalance the hoop force.

The unknown variables are expanded order by order using the inverse aspect ratio r/R as the small parameter ϵ, e.g. $\mathbf{B} = \mathbf{B}_0 + \epsilon \mathbf{B}_1$, with the perturbed quantities labeled by the subscript '1'. The external vertical magnetic field $\mathbf{B}_z$ is assumed to be of order ϵ. Therefore, $\mathbf{B}_1$ contains both the perturbed magnetic field from the toroidal effects and the external vertical magnetic field. The quasi-toroidal coordinate system (r, θ, ϕ) in figure 1.1 is used. The lowest order equilibrium is determined using the following equation:

$$\frac{\partial}{\partial r}\left(\frac{B_\phi^2 + B_\theta^2}{2} + \mu_0 P\right) + \frac{B_\theta^2}{r} = 0. \tag{1.26}$$

To simplify the notation, we drop the subscript 0 for lowest order quantities. We note that here (1.26) is not the usual cylinder solution, as given in [7]. As shown below, the vertical magnetic field can lead to a magnetic surface shift in the cylinder limit $R \to \infty$.

The magnetic surfaces are assumed to be shifted circles and are denoted by $r' = r - \epsilon\xi_1$, where ξ_1 is introduced to describe the surface shift. Noting that the plasma pressure is a surface function $P(r')$, one can obtain [7]

$$P_1(r, \theta) = -\xi_1(r, \theta)\frac{\mathrm{d}P}{\mathrm{d}r}. \tag{1.27}$$

The first order quantities are expanded in the Fourier series and, for the current problem, keeping the first Fourier harmonic is sufficient, e.g. $\xi_1 = \Re e\,\{\xi_1 \exp\{\mathrm{i}\theta\}\}$. Without causing ambiguity, we have used the same symbol for the first Fourier component as the original.

The linearized equations for the first Fourier components can be obtained as follows:

$$-\frac{\partial^2 P}{\partial r^2}\xi_1 - \frac{\partial P}{\partial r}\frac{\partial \xi_1}{\partial r} = j_\theta B_{\phi 1} - j_\phi B_{\theta 1} + j_{\theta 1} B_\phi - j_{\phi 1} B_\theta, \tag{1.28}$$

$$B_{r1} = \mathrm{i}\frac{\xi_1}{r} B_\theta - \mathrm{i} B_z \tag{1.29}$$

$$B_{\theta 1} = -\frac{r}{R} B_\theta - \frac{\partial \xi_1 B_\theta}{\partial r} + B_z, \tag{1.30}$$

$$B_{\phi 1} = -\frac{r}{R} B_\phi + \mu_0 \xi_1 j_\theta, \tag{1.31}$$

$$j_{r1} = \mathrm{i}\frac{\xi_1}{r} j_\theta, \tag{1.32}$$

$$j_{\theta 1} = -\frac{r}{R} j_\theta - \frac{\partial \xi_1 j_\theta}{\partial r}, \tag{1.33}$$

$$\mu_0 j_{\phi 1} = \frac{1}{r}\frac{\partial r B_{r\theta 1}}{\partial r} - \mathrm{i}\frac{1}{r} B_{r1}. \tag{1.34}$$

Here, (1.28) is obtained from the radial projection of the force balance equation (1.7) and, respectively, (1.29) and (1.32) are obtained from $\mathbf{B} \cdot \nabla P = 0$ and $\mathbf{j} \cdot \nabla P = 0$, (1.30) and (1.33) are obtained from $\nabla \cdot \mathbf{B} = 0$ and $\nabla \cdot \mathbf{j} = 0$, and (1.31) and (1.34) are obtained from the radial and toroidal components of Ampere's law in (1.8).

Inserting (1.29)–(1.34) into (1.28), one obtains the first order equilibrium equation

$$\frac{1}{r}\frac{\partial}{\partial r}\left[r B_\theta^2 \frac{\partial \xi_1}{\partial r} \right] + \frac{1}{R} B_\theta^2 - 2\mu_0 \frac{r}{R}\frac{\partial P}{\partial r} - \mu_0 j_\phi B_z = 0. \tag{1.35}$$

This equation can be used to solve for the magnetic surface shift ξ_1 as soon as the two equilibrium profiles are specified, e.g. $B_\theta(r)$ and $P(r)$, together with the external vertical magnetic field B_z.

From (1.35) one can find the force balance picture as follows: $B_\theta(r)$ and $P(r)$ in the second and third terms cause the magnetic surface to shift outward from the tokamak axisymmetric axis. However, the $-j_\phi B_z$ in the final term can push back the plasma torus toward the center. In the cylinder limit $R \to \infty$ the second and third terms in (1.35) vanish; the magnetic surface shift in this case is determined only by the external vertical magnetic field B_z.

Note that both plasma current and vertical magnetic field contribute to the poloidal magnetic field B_θ. Therefore, B_θ^2 in the first term of (1.35) becomes $B_\theta^{p^2} + B_z^2/2$, where B_θ^p represents the poloidal field induced by plasma current. At the small r limit, one has $B_\theta^p \propto r$, but B_z remains finite. Taking into account the $B_z^2/2$ contribution, one can find that the vertical magnetic field contribution to the magnetic surface shift ξ_1 (the last term in (1.35)) converges at the $r \to 0$ limit.

Bibliography

[1] Bernstein I B, Frieman E A, Kruskal M D and Kulsrud R M 1958 An energy principle for hydromagnetic stability problems *Proc. R. Soc.* A **244** 17–40

[2] Grad H and Rubin H 1958 Hydromagnetic equilibria and force-free fields *Proc. 2nd UN Conf. on the Peaceful Uses of Atomic Energy* vol 31 (Geneva: International Atomic Energy Agency) p 190

[3] Shafranov V D 1966 Plasma equilibrium in a magnetic field *Reviews of Plasma Physics* vol 2 (New York: Consultants Bureau) p 103

[4] Grimm R C, Greene J M and Johnson J L 1976 *Methods of Computational Physics* (New York: Academic)

[5] Hamada S 1962 Hydromagnetic equilibria and their proper coordinates *Nucl. Fusion* **2** 23

[6] Boozer A H 1982 Establishment of magnetic coordinates for a given magnetic field *Phys. Fluids* **25** 520–1

[7] Solov'ev L S and Shafranov V D 1970 Plasma confinement in closed magnetic systems *Reviews of Plasma Physics* vol 5, ed M A Leontovich (New York: Consultants Bureau) p 1

Chapter 2

Ideal MHD instabilities

In this chapter, we will describe the linear ideal MHD stability theory. We mainly focus on the theory in toroidally confined plasmas, as reviewed in [1]. Although many-charged particle systems, with long mean free paths and long range correlations, are intrinsically complicated, MHD theory is relatively simple and gives rise to relevant theoretical predictions for experiments. Currently, tokamaks are designed primarily according to the MHD equilibrium and stability theories. The magnetic confinement of fusion plasmas would not be expected survive if the MHD theory predicted major instabilities.

In conventional fluid theory, particle collisions are the cause of particle localization. However, for magnetically confined fusion plasmas the collision frequency is usually low; particle collisions alone are insufficient to hold particles together. The relevance of the partial fluid description of magnetically confined fusion plasmas relies on the presence of a strong magnetic field. Charged particles are tied to magnetic field lines by gyromotions. Therefore, in the direction perpendicular to the magnetic field lines, the magnetic field can play the role of localization, so the perpendicular MHD description becomes relevant at least in the lowest order. One would expect that the perpendicular MHD description needs to be modified only when the effects of the finite Larmor radius (FLR) become significant.

In the direction parallel to the magnetic field, however, particles can move quite freely. The collision frequency is not high enough to hold charged particles together to establish a local thermal equilibrium. One cannot define the local thermal parameters, such as the parallel fluid velocity, density, temperature, etc. The trapped particle effect, wave–particle resonances and parallel electric field effects need to be included. Plasma behavior in the parallel direction is intrinsically non-fluid and requires a kinetic description. Surprisingly, even under these circumstances the MHD description yields valuable and relevant theoretical predictions without major modifications in the relevant low ($\omega \ll \omega_{si}$) and intermediate ($\omega_{si} \ll \omega \ll \omega_{se}$) frequency regimes, where ω is the mode frequency, and ω_{si} and ω_{se} represent, respectively, the ion and electron acoustic frequencies. In the low frequency regime the coupling of parallel motion results only in an enhanced apparent mass effect [2],

doi:10.1088/978-1-6270-5423-2ch2

while in the intermediate frequency regime the kinetic effects only give rise to a new phenomenological ratio of special heats in the leading order [3].

The basic set of MHD equations are given in (1.1)–(1.6). The linearization of this set of equations gives [4]

$$-\rho_m \omega^2 \boldsymbol{\xi} = \delta \mathbf{J} \times \mathbf{B} + \mathbf{J} \times \delta \mathbf{B} - \nabla \delta P, \tag{2.1}$$

$$\delta \mathbf{B} = \nabla \times \boldsymbol{\xi} \times \mathbf{B}, \tag{2.2}$$

$$\mu_0 \delta \mathbf{J} = \nabla \times \delta \mathbf{B}, \tag{2.3}$$

$$\delta P = -\boldsymbol{\xi} \cdot \nabla P - \Gamma P \nabla \cdot \boldsymbol{\xi}, \tag{2.4}$$

where $\boldsymbol{\xi} = \mathbf{v}/(-i\omega)$ represents the plasma displacement and the time dependence of the perturbed quantities is assumed to be of exponential type $\exp\{-i\omega t\}$. Inserting (2.2)–(2.4) into (2.1), one obtains a single equation for $\boldsymbol{\xi}$:

$$\begin{aligned} -\rho_m \omega^2 \boldsymbol{\xi} = & \frac{1}{\mu_0} \nabla \times (\nabla \times \boldsymbol{\xi} \times \mathbf{B}) \times \mathbf{B} + \mathbf{J} \times \nabla \times \boldsymbol{\xi} \times \mathbf{B} \\ & + \nabla(\boldsymbol{\xi} \cdot \nabla P + \Gamma P \nabla \cdot \boldsymbol{\xi}). \end{aligned} \tag{2.5}$$

We have not included toroidal rotation effects in the linearized equations (2.1)–(2.4). For most tokamak experiments rotation is subsonic, i.e. the rotation speed is much smaller than the ion thermal speed. In this case the centrifugal and Coriolis forces from plasma rotation are smaller than the effects from particle thermal motion—the plasma pressure effect. Therefore, the rotation effects can be taken into account simply by introducing the Doppler frequency shift: $\omega \to \omega + n\Omega_{\rm rot}$ in the MHD equation (2.5), where $\Omega_{\rm rot}$ is the toroidal rotation frequency and n denotes the toroidal mode number [5, 6].

There are three fundamental types of waves in magnetic confined plasmas: the compressional Alfvén wave, the shear Alfvén wave and the parallel acoustic wave. The compressional Alfvén wave characterizes the wave due to the perpendicular compression and restoration of the magnetic field. It mainly propagates in the direction perpendicular to the magnetic field. Since plasmas are frozen in the magnetic field, such a magnetic field compression also induces plasma compression. Note that the ratio of plasma pressure to magnetic pressure (referred to as plasma beta β) is usually low. The compression and restoration forces mainly result from the energy of the magnetic field.

The shear Alfvén wave describes the oscillation due to magnetic field line bending and restoration. It mainly propagates along the magnetic field lines. Since a longer wavelength is allowed, the frequency (or restoration force) of shear Alfvén waves is usually lower than that of the compressional Alfvén waves by an order of $\lambda_\perp/\lambda_\parallel$—the ratio of the perpendicular to parallel wavelengths. Therefore, the shear Alfvén wave is often coupled to plasma instabilities.

Another fundamental wave in magnetic confined plasmas is the parallel acoustic wave (sound wave). Since plasma can move freely along the magnetic field lines without being affected by the Lorentz force, parallel acoustic waves can prevail in plasmas. The various types of electrostatic drift waves are related to the parallel acoustic wave. Due to the low beta assumption, one can prove that the frequency of the ion acoustic waves is lower than that of the shear Alfvén wave by an order of $\sqrt{\beta}$.

The behaviors of these three waves in slab geometry have been widely discussed in many MHD publications. In toroidal geometry, however, the MHD equation (2.5) can be difficult to deal with. One usually needs to separate the time scales for the three fundamental waves to reduce the problem. This scale separation is realized through proper projections and reduction of the MHD equation (2.5).

There are three projections for the MHD equation (2.5). We introduce three unit vectors, $\mathbf{e}_b = \mathbf{B}/B$, $\mathbf{e}_1 = \nabla\psi/|\nabla\psi|$ and $\mathbf{e}_2 = \mathbf{e}_b \times \mathbf{e}_1$, for these projections. The $\mathbf{e}_2$ projection of the MHD equation (2.5) gives

$$\begin{aligned}\mathbf{e}_1 \cdot \nabla \times \delta\mathbf{B} = &-\frac{gP'}{B^2}\mathbf{e}_1 \cdot \delta\mathbf{B} - g'\mathbf{e}_1 \cdot \delta\mathbf{B} + \frac{1}{B}\mathbf{e}_2 \cdot \nabla\big(|\nabla P|\, \mathbf{e}_1 \cdot \boldsymbol{\xi}\big)\\ &+ \Gamma P\frac{1}{B}\mathbf{e}_2 \cdot \nabla(\nabla \cdot \boldsymbol{\xi}) + \frac{\rho_m\omega^2}{B}\mathbf{e}_2 \cdot \boldsymbol{\xi}.\end{aligned} \tag{2.6}$$

Similarly, the $\mathbf{e}_1$ projection of the MHD equation (2.5) yields

$$\begin{aligned}\mathbf{e}_2 \cdot \nabla \times \delta\mathbf{B} = &-\frac{gP'}{B^2}\mathbf{e}_2 \cdot \delta\mathbf{B} - g'\mathbf{e}_2 \cdot \delta\mathbf{B} - \frac{P'|\nabla\psi|}{B^2}\mathbf{e}_b \cdot \delta\mathbf{B} - \frac{1}{B}\mathbf{e}_1 \cdot \nabla\big(|\nabla P|\, \mathbf{e}_1 \cdot \boldsymbol{\xi}\big)\\ &- \Gamma P\frac{1}{B}\mathbf{e}_1 \cdot \nabla(\nabla \cdot \boldsymbol{\xi}) - \frac{\rho_m\omega^2}{B}\mathbf{e}_1 \cdot \boldsymbol{\xi}.\end{aligned} \tag{2.7}$$

The $\mathbf{e}_b$ projection of the MHD equation (2.5) can be reduced to, using $\nabla \cdot \boldsymbol{\xi}$ as an independent unknown [7],

$$\Gamma P\mathbf{B} \cdot \nabla\left(\frac{1}{B^2}\mathbf{B} \cdot \nabla\nabla \cdot \boldsymbol{\xi}\right) + \rho_m\omega^2\nabla \cdot \boldsymbol{\xi} = \rho_m\omega^2\nabla \cdot \boldsymbol{\xi}_\perp. \tag{2.8}$$

Noting that $\delta\mathbf{J}$ and $\delta\mathbf{B}$ are completely determined by $\boldsymbol{\xi}_\perp$, one can see that the set of equations (2.6)–(2.8) is complete for determining the two components of $\boldsymbol{\xi}_\perp$ and the scalar unknown $\nabla \cdot \boldsymbol{\xi}$.

The two perpendicular equations of motion, (2.6) and (2.7), result from perpendicular projections of the MHD equation (2.1) and therefore contain the compression/restoration force from the excitation of the compressional Alfvén wave. Note that the compressional Alfvén wave results from the term $\delta\mathbf{J} \times \mathbf{B} + \mathbf{J} \times \delta\mathbf{B} + \nabla\delta P \to \nabla(\mathbf{B} \cdot \delta\mathbf{B} + \delta P)$ in (2.1). Therefore the curl operation in deriving (2.9) can suppress the compressional Alfvén wave. Applying the operator $\nabla \cdot (\mathbf{B}/B^2) \times (\cdots)$ on (2.1), one obtains

$$\begin{aligned}\nabla \cdot \frac{\mathbf{B}}{B^2} \times \rho_m\omega^2\boldsymbol{\xi} = &\ \mathbf{B} \cdot \nabla\frac{\mathbf{B} \cdot \delta\mathbf{J}}{B^2} + \delta\mathbf{B} \cdot \nabla\sigma - \mathbf{J} \cdot \nabla\frac{\mathbf{B} \cdot \delta\mathbf{B}}{B^2}\\ &+ \nabla \times \frac{\mathbf{B}}{B^2} \cdot \nabla\delta P,\end{aligned} \tag{2.9}$$

where $\sigma = \mathbf{J} \cdot \mathbf{B}/B^2$. Equation (2.9) is often referred to as the shear Alfvén law or vorticity equation.

Equations (2.9), (2.6) and (2.8) characterize, respectively, the three fundamental MHD waves: the shear Alfvén, compressional Alfvén and parallel acoustic waves. In the following sections we will first describe the global MHD spectrum in tokamaks. Then, we will individually examine the interchange modes, ballooning modes, peeling and free boundary ballooning modes, toroidal Alfvén eigenmodes (TAEs) and kinetically driven MHD modes.

2.1 The global MHD spectrum

In this section we describe the global MHD mode spectrum in tokamaks. We first discuss the complicated global MHD theory, since it can draw an overall theoretical picture of the various localized mode theories to be described later in this chapter. Discussion of the MHD spectrum can help in understanding the role of the three fundamental waves in stability theories. Due to the complexity of global MHD analyses, they have to be addressed numerically. Nevertheless, one can skip the complicated details of how the matrices are constructed and the eigenmodes computed to understand the fundamental theoretical framework.

Several excellent numerical codes have been developed in this field to study the linear MHD stability of toroidally confined plasmas, such as the PEST [8, 9], GATO [10], DCON [11] and AEGIS [12] codes, etc. In this section we focus on describing the AEGIS code formalism, in view of the fact that AEGIS is an adaptive MHD shooting code and is capable of studying the MHD continuum [13].

Let us first describe the toroidal system to be investigated. The core part is the plasma torus, which is surround by a resistive wall. Between the plasma torus and resistive wall is the inner vacuum region and outside the resistive wall is the outer vacuum region, which extends to infinity. For simplicity, it is assumed that the wall is thin. We denote the interfaces between the plasma torus and inner vacuum region, the inner vacuum region and wall, and the wall and outer vacuum region as ψ_a, ψ_{b-} and ψ_{b+}, respectively.

The initial equation is the single-fluid MHD equation (2.5). This is a vector equation and can be projected onto three directions to obtain scalar equations. The parallel projection has been derived in (2.8). From the parallel equation one can solve for $\nabla \cdot \boldsymbol{\xi}$, which is the only unknown describing the parallel motion in the perpendicular momentum equations. In principle, the parallel motion can not be described by the MHD model, since particles are not localized along magnetic field lines. The AEGIS-K code has been developed to include the parallel dynamics using a kinetic description [14].

Using the general flux coordinates in (1.19), the magnetic field line displacement can be decomposed as follows

$$\boldsymbol{\xi} \times \mathbf{B} = \xi_s \nabla\psi + \xi_\psi \chi'(\nabla\zeta - q\nabla\theta). \tag{2.10}$$

Since we are dealing with a linear problem, Fourier transformation can be used to decompose the perturbed quantities in the poloidal and toroidal directions,

$$\xi \exp\{-in\zeta\} = \sum_{m=-\infty}^{\infty} \xi_m \frac{1}{\sqrt{2\pi}} \exp\{i(m\theta - n\zeta)\}. \tag{2.11}$$

Here, $\xi_m = \int_{-\pi}^{\pi} d\theta\xi \exp\{-im\theta\}/\sqrt{2\pi}$. With the assumption of toroidal symmetry, only a single toroidal Fourier component needs to be considered. As usual, the equilibrium quantities can be decomposed as matrices in the poloidal Fourier space, for example

$$\mathcal{J}_{mm'} = \frac{1}{2\pi}\int_{-\pi}^{\pi} d\theta J(\theta) e^{i(m'-m)\theta}.$$

In the poloidal Fourier decomposition, the Fourier components are cut off from both the lower and upper sides by $m_{\min}$ and $m_{\max}$, respectively. Therefore, the total number of Fourier components under consideration is $M = m_{\max} - m_{\min} + 1$. We use bold face (or alternatively $[[\cdots]]$) to represent the Fourier space vectors, and calligraphic capital letters (or alternatively $\langle\cdots\rangle$) to represent the corresponding equilibrium matrices (e.g. $\mathcal{J}$ for J) in the poloidal Fourier space.

We will express the matrix equations to be derived in a normalized form. To achieve this, the perpendicular MHD equation (2.5) is normalized by multiplying by the factor $\mu_0 R_0/(\rho_{m0} B_0^2)$ and the parallel momentum equation (2.8) is normalized by dividing by $\rho_{m0}\omega_{A0}^2$, where the subscript '0' indicates the value at the magnetic axis. In this normalization the length, including $\boldsymbol{\xi}$, is normalized by the major radius R_0, P becomes the beta value $\mu_0 P/B_0^2$, the mass density ρ_m becomes ρ_m/ρ_{m0} and the frequency is normalized by the Alfvén frequency ω/ω_{A0}, where $\omega_{A0} = B_0/(\sqrt{\mu_0\rho_{m0}}R_0)$.

We first reduce the parallel momentum equation (2.8) to a matrix equation. One can prove that [24]

$$\nabla\cdot\boldsymbol{\xi}_\perp + 2\boldsymbol{\kappa}\cdot\boldsymbol{\xi} = \frac{\mu_0\boldsymbol{\xi}\cdot\nabla P}{B^2} - \frac{\mathbf{B}\cdot\delta\mathbf{B}}{B^2}.$$

Using this identity to express $\nabla\cdot\boldsymbol{\xi}_\perp$, equation (2.8) can be reduced to

$$\mathcal{L}_s[[\nabla\cdot\boldsymbol{\xi}]] = \mathcal{R}_s\boldsymbol{\xi}_s + \mathcal{R}_s\boldsymbol{\xi}'_\psi + \mathcal{R}_s\boldsymbol{\xi}_\psi. \tag{2.12}$$

where

$$\mathcal{L}_s = -\Gamma P\chi'^2\left\langle\frac{1}{J}\right\rangle \mathcal{Q}\left\langle\frac{1}{B^2 J}\right\rangle \mathcal{Q} + \rho_m\omega^2, \tag{2.13}$$

$$\mathcal{R}_s = i\rho_m\omega^2\chi'\left\langle\frac{1}{JB^2}\right\rangle\left[n(\mathcal{G}_{22} + q\mathcal{G}_{23}) + (\mathcal{G}_{23} + q\mathcal{G}_{33})\mathcal{M}\right] - 2\mathcal{K}_s, \tag{2.14}$$

$$\mathcal{R}_1 = \rho_m\omega^2\chi'^2\left\langle\frac{1}{JB^2}\right\rangle\left[(\mathcal{G}_{22} + q\mathcal{G}_{23}) + q(\mathcal{G}_{23} + q\mathcal{G}_{33})\right], \tag{2.15}$$

$$\mathcal{R}_0 = -\rho_m\omega^2\chi'\left\langle\frac{1}{JB^2}\right\rangle\left[i\chi'(\mathcal{G}_{12} + q\mathcal{G}_{31})\mathcal{Q} - \chi''(\mathcal{G}_{22} + q\mathcal{G}_{23}) - (q\chi')'(\mathcal{G}_{23} + q\mathcal{G}_{33})\right]$$
$$+ P'\left\langle\frac{1}{B^2}\right\rangle - 2\mathcal{K}_\psi. \tag{2.16}$$

Here, $\mathcal{K}_s = \langle \mathbf{B} \times \mathbf{s} \cdot \boldsymbol{\kappa}/B^2 \rangle$, $\mathcal{K}_\psi = \langle \mathbf{B} \times \nabla\psi \cdot \boldsymbol{\kappa}/B^2 \rangle$, $\mathcal{M}_{mm'} = m\mathcal{I}_{mm'}$, $\mathcal{Q}_{mm'} = (m - nq)\mathcal{I}_{mm'}$, $\mathcal{I}$ is the unit matrix, $\mathbf{s} = \chi'(\nabla\zeta - q\nabla\theta)$, and the following metric matrices are also introduced

$$\mathcal{G}_{11} = \langle J(\nabla\theta \times \nabla\zeta) \cdot (\nabla\theta \times \nabla\zeta) \rangle,$$

$$\mathcal{G}_{22} = \langle J(\nabla\zeta \times \nabla\psi) \cdot (\nabla\zeta \times \nabla\psi) \rangle,$$

$$\mathcal{G}_{33} = \langle J(\nabla\psi \times \nabla\theta) \cdot (\nabla\psi \times \nabla\theta) \rangle,$$

$$\mathcal{G}_{12} = \langle J(\nabla\theta \times \nabla\zeta) \cdot (\nabla\zeta \times \nabla\psi) \rangle,$$

$$\mathcal{G}_{31} = \langle J(\nabla\psi \times \nabla\theta) \cdot (\nabla\theta \times \nabla\zeta) \rangle,$$

$$\mathcal{G}_{23} = \langle J(\nabla\zeta \times \nabla\psi) \cdot (\nabla\psi \times \nabla\theta) \rangle.$$

The equation of parallel motion, (2.12), can be solved by matrix inversion

$$[[\nabla \cdot \boldsymbol{\xi}]] = \mathcal{L}_s^{-1}\left(\mathcal{R}_s\boldsymbol{\xi}_s + \mathcal{R}_s\boldsymbol{\xi}'_\psi + \mathcal{R}_s\boldsymbol{\xi}_\psi\right). \tag{2.17}$$

Next, we project (2.5) onto the two directions $J^2\nabla\theta \times \nabla\zeta \cdot \mathbf{B} \times [\cdots \times \mathbf{B}]/B^2$ and $(1/q\chi')J^2\nabla\zeta \times \nabla\psi \cdot \mathbf{B} \times [\cdots \times \mathbf{B}]/B^2$, and then introduce the Fourier transformation in (2.11) to the two projected equations. These procedures lead to the following set of differential equations in matrices

$$\left(\mathcal{B}^{\mathrm{L}}\xi_s + \mathcal{D}\boldsymbol{\xi}'_\psi + \mathcal{E}\boldsymbol{\xi}_\psi\right)' - \left(\mathcal{C}^{\mathrm{L}}\boldsymbol{\xi}_s + \mathcal{E}^{\mathrm{L}}\boldsymbol{\xi}'_\psi + \mathcal{H}\boldsymbol{\xi}_\psi\right) = 0, \tag{2.18}$$

$$\mathcal{A}\boldsymbol{\xi}_s + \mathcal{B}\boldsymbol{\xi}'_\psi + \mathcal{C}\boldsymbol{\xi}_\psi = 0. \tag{2.19}$$

Here, we have used the solution of parallel equation in (2.17) to express the plasma compressibility term $[[\nabla \cdot \boldsymbol{\xi}]]$ in (2.5). The equilibrium matrices then contain three contributions: the incompressible plasma responses, plasma compressibility effects and inertia contributions, e.g. $\mathcal{A} = \mathcal{A}_p + \mathcal{A}_c + \gamma^2\mathcal{A}_i$, where

$$\mathcal{A}_p = n\left(n\mathcal{G}_{22} + \mathcal{G}_{23}\mathcal{M}\right) + \mathcal{M}\left(n\mathcal{G}_{23} + \mathcal{G}_{33}\mathcal{M}\right),$$

$$\mathcal{B}_p = -i\chi'\left[n\left(\mathcal{G}_{22} + q\mathcal{G}_{23}\right) + \mathcal{M}\left(\mathcal{G}_{23} + q\mathcal{G}_{33}\right)\right],$$

$$\mathcal{C}_p = -i\left[\chi''\left(n\mathcal{G}_{22} + \mathcal{M}\mathcal{G}_{23}\right) + (q\chi')'\left(n\mathcal{G}_{23} + \mathcal{M}\mathcal{G}_{33}\right)\right]$$
$$- \chi'\left(n\mathcal{G}_{12} + \mathcal{M}\mathcal{G}_{31}\right)\mathcal{Q} + i\left(g'\mathcal{Q} - \mu_0 nP'\mathcal{J}/\chi'\right),$$

$$\mathcal{D}_p = \chi'^2\left[\left(\mathcal{G}_{22} + q\mathcal{G}_{23}\right) + q\left(\mathcal{G}_{23} + q\mathcal{G}_{33}\right)\right],$$

$$\mathcal{E}_p = \chi'\left[\chi''\left(\mathcal{G}_{22} + q\mathcal{G}_{23}\right) + (q\chi')'\left(\mathcal{G}_{23} + q\mathcal{G}_{33}\right)\right] - i\chi'^2\left(\mathcal{G}_{12} + q\mathcal{G}_{31}\right)\mathcal{Q} + \mu_0 P'\mathcal{J},$$

$$\mathcal{H}_{\mathrm{p}} = \chi''\left[\chi''\mathcal{G}_{22} + (q\chi')'\mathcal{G}_{23}\right] + (q\chi')'\left[\chi''\mathcal{G}_{23} + (q\chi')'\mathcal{G}_{33}\right]$$

$$+ \mathrm{i}\chi'\left[\chi''\left(\mathcal{M}\mathcal{G}_{12} - \mathcal{G}_{12}\mathcal{M}\right) + (q\chi')'\left(\mathcal{M}\mathcal{G}_{31} - \mathcal{G}_{31}\mathcal{M}\right)\right]$$

$$+ \chi'^2\mathcal{Q}\mathcal{G}_{11}\mathcal{Q} + \mu_0 P'\chi''\mathcal{J}/\chi' + \mu_0 P'\mathcal{J}' - g'q'\chi'\mathcal{I},$$

$$\mathcal{A}_{\mathrm{c}} = \mathrm{i}\Gamma P\left\{\frac{\mathcal{J}}{q\chi'}\mathcal{M} - \frac{\chi'}{q}\left\langle\frac{1}{B^2}\right\rangle(\mathcal{G}_{22} + q\mathcal{G}_{23})\mathcal{Q}\right\}\mathcal{L}_{\mathrm{s}}^{-1}\mathcal{R}_{\mathrm{s}},$$

$$\mathcal{B}_{\mathrm{c}} = \mathrm{i}\Gamma P\left\{\frac{\mathcal{J}}{q\chi'}\mathcal{M} - \frac{\chi'}{q}\left\langle\frac{1}{B^2}\right\rangle(\mathcal{G}_{22} + q\mathcal{G}_{23})\mathcal{Q}\right\}\mathcal{L}_{\mathrm{s}}^{-1}\mathcal{R}_{1},$$

$$\mathcal{C}_{\mathrm{c}} = \mathrm{i}\Gamma P\left\{\frac{\mathcal{J}}{q\chi'}\mathcal{M} - \frac{\chi'}{q}\left\langle\frac{1}{B^2}\right\rangle(\mathcal{G}_{22} + q\mathcal{G}_{23})\mathcal{Q}\right\}\mathcal{L}_{\mathrm{s}}^{-1}\mathcal{R}_{0},$$

$$\mathcal{B}_{\mathrm{c}}^{\mathrm{L}} = -\Gamma P\mathcal{J}\mathcal{L}_{\mathrm{s}}^{-1}\mathcal{R}_{\mathrm{s}},$$

$$\mathcal{D}_{\mathrm{c}} = -\Gamma P\mathcal{J}\mathcal{L}_{\mathrm{s}}^{-1}\mathcal{R}_{1},$$

$$\mathcal{E}_{\mathrm{c}} = -\Gamma P\mathcal{J}\mathcal{L}_{\mathrm{s}}^{-1}\mathcal{R}_{0},$$

$$\mathcal{C}_{\mathrm{c}}^{\mathrm{L}} = \Gamma P\left[\mathrm{i}\chi'^2\left\langle\frac{1}{B^2}\right\rangle(\mathcal{G}_{12} + q\mathcal{G}_{31})\mathcal{Q} + \mathcal{J}'\right]\mathcal{L}_{\mathrm{s}}^{-1}\mathcal{R}_{\mathrm{s}},$$

$$\mathcal{E}_{\mathrm{c}}^{\mathrm{L}} = \Gamma P\left[\mathrm{i}\chi'^2\left\langle\frac{1}{B^2}\right\rangle(\mathcal{G}_{12} + q\mathcal{G}_{31})\mathcal{Q} + \mathcal{J}'\right]\mathcal{L}_{\mathrm{s}}^{-1}\mathcal{R}_{1},$$

$$\mathcal{H}_{\mathrm{c}} = \Gamma P\left[\mathrm{i}\chi'^2\left\langle\frac{1}{B^2}\right\rangle(\mathcal{G}_{12} + q\mathcal{G}_{31})\mathcal{Q} + \mathcal{J}'\right]\mathcal{L}_{\mathrm{s}}^{-1}\mathcal{R}_{0},$$

$$\mathcal{A}_{\mathrm{i}} = \frac{B_0^2}{R_0^2}\left\langle J\frac{\rho_m}{B^2}\,|\nabla\psi|^2\right\rangle,$$

$$\mathcal{C}_{\mathrm{i}} = \frac{B_0^2}{R_0^2}\left\langle\chi' J\frac{\rho_m}{B^2}(\nabla\psi\cdot\nabla\zeta - q\nabla\psi\cdot\nabla\theta)\right\rangle,$$

$$\mathcal{H}_{\mathrm{i}} = \frac{B_0^2}{R_0^2}\left\langle\chi'^2 J\frac{\rho_m}{B^2}\left(|\nabla\zeta|^2 + q^2\,|\nabla\theta|^2 - 2\nabla\theta\cdot\nabla\zeta\right)\right\rangle,$$

and $\mathcal{B}_{\mathrm{i}} = \mathcal{D}_{\mathrm{i}} = \mathcal{E}_{\mathrm{i}} = 0$, γ denotes the normalized growth rate. Here, the incompressible plasma matrices and inertia matrices are self-adjoint, for example $\mathcal{B}_{\mathrm{p}}^{\mathrm{L}} = \mathcal{B}_{\mathrm{p}}^{\dagger}$, etc.

We can reduce the set of equations (2.18) and (2.19) into a set of second order differential equations as in the DCON formalism [11]. By solving (2.19), one obtains

$$\xi_s = -\mathcal{A}^{-1}\mathcal{B}\xi'_\psi - \mathcal{A}^{-1}\mathcal{C}\xi_\psi.$$

Inserting this solution into (2.18), we obtain a single matrix equation governing the ideal MHD modes:

$$\frac{\mathrm{d}}{\mathrm{d}\psi}(\mathcal{F}\xi' + \mathcal{K}\xi) - \left(\mathcal{K}^{\mathrm{L}}\xi' + \mathcal{G}\xi\right) = 0, \tag{2.20}$$

where the subscript ψ for $\boldsymbol{\xi}$ has been dropped and

$$\begin{aligned}
\mathcal{F} &= \mathcal{D} - \mathcal{B}^{\mathrm{L}}\mathcal{A}^{-1}\mathcal{B},\\
\mathcal{K} &= \mathcal{E} - \mathcal{B}^{\mathrm{L}}\mathcal{A}^{-1}\mathcal{C},\\
\mathcal{K}^{\mathrm{L}} &= \mathcal{E}^{\mathrm{L}} - \mathcal{C}^{\mathrm{L}}\mathcal{A}^{-1}\mathcal{B},\\
\mathcal{G} &= \mathcal{H} - \mathcal{C}^{\mathrm{L}}\mathcal{A}^{-1}\mathcal{C}.
\end{aligned}$$

These matrices can be further simplified as follows

$$\begin{aligned}
\mathcal{F} = \frac{\chi'^2}{n^2}\Bigg\{ & Q\mathcal{G}_{33}Q + \gamma^2\mathcal{A}_\mathrm{i} + \left(\mathcal{A}_\mathrm{c} + \mathrm{i}\frac{n}{\chi'}\mathcal{B}_\mathrm{c}^\mathrm{L}\right) + \frac{n^2}{\chi'^2}\left(\mathcal{D}_\mathrm{c} - \mathrm{i}\frac{\chi'}{n}\mathcal{B}_\mathrm{c}\right)\\
& - \left[\gamma^2\mathcal{A}_\mathrm{i} + \left(\mathcal{A}_\mathrm{c} + \mathrm{i}\frac{n}{\chi'}\mathcal{B}_\mathrm{c}^\mathrm{L}\right) + Q\left(n\mathcal{G}_{23} + \mathcal{G}_{33}\mathcal{M}\right)\right]\\
& \times \mathcal{A}^{-1}\left[\gamma^2\mathcal{A}_\mathrm{i} + \left(\mathcal{A}_\mathrm{c} - \mathrm{i}\frac{n}{\chi'}\mathcal{B}_\mathrm{c}\right) + \left(n\mathcal{G}_{23} + \mathcal{M}\mathcal{G}_{33}\right)Q\right]\Bigg\},
\end{aligned} \tag{2.21}$$

$$\begin{aligned}
\mathcal{K} = \frac{\chi'}{n}\Bigg\{ & \mathrm{i}\left[\gamma^2\mathcal{A}_\mathrm{i} + \left(\mathcal{A}_\mathrm{c} + \mathrm{i}\frac{n}{\chi'}\mathcal{B}_\mathrm{c}^\mathrm{L}\right) + Q\left(n\mathcal{G}_{23} + \mathcal{G}_{33}\mathcal{M}\right)\right]\mathcal{A}^{-1}\mathcal{C}\\
& - Q\left[\chi''\mathcal{G}_{23} + (q\chi')'\mathcal{G}_{33} - \mathrm{i}\chi'\mathcal{G}_{31}Q - g'\mathcal{I}\right] - \mathrm{i}\gamma^2\mathcal{C}_\mathrm{i}\Bigg\} + \left(\mathcal{E}_\mathrm{c} - \mathrm{i}\chi'\frac{1}{n}\mathcal{C}_\mathrm{c}\right),
\end{aligned} \tag{2.22}$$

$$\begin{aligned}
\mathcal{K}^\mathrm{L} = \frac{\chi'}{n}\Bigg\{ & -\mathrm{i}\mathcal{C}^\mathrm{L}\mathcal{A}^{-1}\left[\gamma^2\mathcal{A}_\mathrm{i} + \left(\mathcal{A}_\mathrm{c} - \mathrm{i}\frac{n}{\chi'}\mathcal{B}_\mathrm{c}\right) + \left(n\mathcal{G}_{23} + \mathcal{M}\mathcal{G}_{33}\right)Q\right]\\
& - \left[\chi''\mathcal{G}_{23} + (q\chi')'\mathcal{G}_{33} + \mathrm{i}\chi'Q\mathcal{G}_{31} - g'\mathcal{I}\right]Q + \mathrm{i}\gamma^2\mathcal{C}_\mathrm{i}\Bigg\} + \left(\mathcal{E}_\mathrm{c}^\mathrm{L} + \mathrm{i}\chi'\frac{1}{n}\mathcal{C}_\mathrm{c}^\mathrm{L}\right).
\end{aligned} \tag{2.23}$$

Here, the simplification of the MHD part is DCON-like [11]. The plasma compressibility matrices can be further combined as follows

$$\mathcal{A}_c - i\frac{n}{\chi'}\mathcal{B}_c = i\Gamma P\frac{1}{q}\left[\frac{1}{\chi'}\mathcal{J}\mathcal{M} - \chi'\left\langle\frac{1}{B^2}\left(G_{22} + qG_{23}\right)\right\rangle\mathcal{Q}\right]\mathcal{L}_s^{-1}\mathcal{R}_{s1},$$

$$\mathcal{A}_c + i\frac{n}{\chi'}\mathcal{B}_c^L = i\Gamma P\frac{1}{q}\left[\frac{1}{\chi'}\mathcal{J}\mathcal{Q} - \chi'\left\langle\frac{1}{B^2}\left(G_{22} + qG_{23}\right)\right\rangle\mathcal{Q}\right]\mathcal{L}_s^{-1}\mathcal{R}_s,$$

$$\left(\mathcal{A}_c + i\frac{n}{\chi'}\mathcal{B}_c^L\right) + \frac{n^2}{\chi'^2}\left(\mathcal{D}_c - i\frac{\chi'}{n}\mathcal{B}_c\right)$$

$$= i\Gamma P\frac{1}{q}\left[\frac{1}{\chi'}\mathcal{J}\mathcal{Q} - \chi'\left\langle\frac{1}{B^2}\left(G_{22} + qG_{23}\right)\right\rangle\mathcal{Q}\right]\mathcal{L}_s^{-1}\mathcal{R}_{s1},$$

$$\mathcal{E}_c - i\chi'\frac{1}{n}\mathcal{C}_c = \Gamma P\frac{1}{nq}\left[\mathcal{J}\mathcal{Q} - \chi'^2\left\langle\frac{1}{B^2}\left(G_{22} + qG_{23}\right)\right\rangle\mathcal{Q}\right]\mathcal{L}_s^{-1}\mathcal{R}_0,$$

$$\mathcal{E}_c^L + i\chi'\frac{1}{n}\mathcal{C}_c^L = i\Gamma P\frac{\chi'}{n}\left[i\chi'^2\left\langle\frac{1}{B^2}\left(G_{12} + qG_{31}\right)\right\rangle\mathcal{Q} + \mathcal{J}'\right]\mathcal{L}_s^{-1}\mathcal{R}_{s1},$$

where $\mathcal{R}_{s1} = \mathcal{R}_s - i(n/\chi')\langle\mathcal{R}_1\rangle$.

Equations (2.18), (2.19) and (2.12) are obtained from three projections of the MHD equation (2.5), which basically describe the shear Alfvén, compressional Alfvén and parallel sound waves. It should be pointed out that the shear and compressional Alfvén waves are coupled in (2.18) and (2.19). The compressional Alfvén mode can be decoupled by constructing the vorticity equation (2.9). Alternatively, by solving (2.19) for $\boldsymbol{\xi}_s$ and inserting it into (2.18) one can also obtain the decoupled shear Alfvén mode equation, (2.20).

With (2.18), (2.19) and (2.12) one can study the MHD mode spectrum. The MHD spectrum has been investigated, for example, in [15] and [16], although only the shear Alfvén and sound wave spectra are detailed. Figure 2.1 shows the spectrum for the case of a circular cross section. The current set of equations (2.18), (2.19) and (2.12) show that there are three sets of spectra corresponding to the three fundamental MHD waves. In deriving the equations describing the shear and compressional Alfvén waves, (2.18) and (2.19), one has to solve the sound wave equation (2.12). The solubility condition for the sound wave equation det $|\mathcal{L}_s| = 0$ in the (ω^2, ψ) space determines exactly the sound spectrum as given in figure 2.1. To derive the shear Alfvén mode equation (2.20), one needs the solution of (2.19). The solubility condition for (2.19) det $|\mathcal{A}| = 0$ gives rise to the compressional Alfvén spectrum. Since the frequency of the compressional Alfvén spectrum is high, its spectrum is above the upper limit of figure 2.1. Finally, the solubility condition for the shear Alfvén mode equation (2.20) det $|\mathcal{F}| = 0$ determines the shear Alfvén spectrum in figure 2.1. The dashed curves for $m = 1$ and 2 are the cylindrical limit, while the solid curves represent the toroidal results.

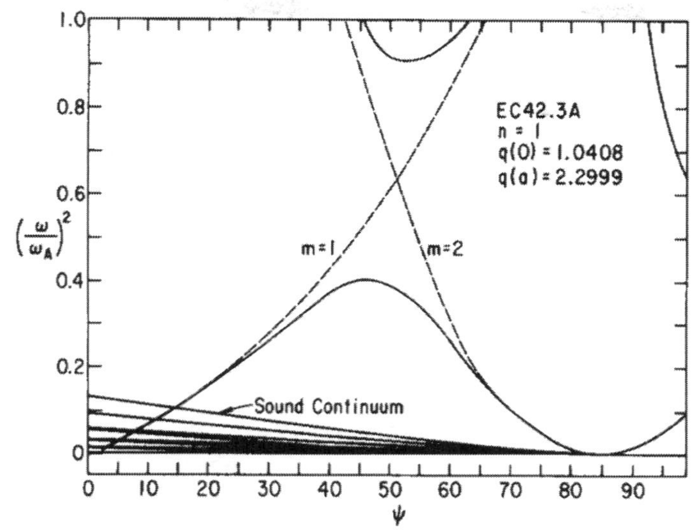

Figure 2.1. The continuum spectrum for $n = 1$ modes in a tokamak with a circular cross section. (Reproduced with permission from [15]. Copyright 1986, AIP Publishing LLC.)

Now let us discuss the connection of the current global theory with the localized analytical theories described later in this chapter. The toroidal interchange mode theory in section 2.3 is derived by employing the mode localization assumption. In this case only two side bands of the mode couplings are considered. The general eigenmode equation (2.20) in the plasma region contains all the side band couplings. Noting that $Q \propto x$, one can see from (2.21)–(2.24) that $\mathcal{F} \propto x^2$ and $\mathcal{K} \propto x$ at the marginal stability $\omega = 0$. We can therefore see the root of the singular layer equation (2.61) in (2.20). For the ballooning mode theory in section 2.4, the ballooning invariance in (2.73) is introduced, so that the set of matrix equation (2.20) can be transformed into an one-dimensional ballooning equation. The TAE theory in section 2.6 mainly uses two Fourier components to construct the eigenmodes.

2.2 Internal and external global MHD modes

In the previous section, we reduced the linear MHD equation (2.5) into a single matrix equation (2.20). In this section, we describe how to obtain the numerical solution of this matrix equation (2.20). In particular, the internal and external global MHD modes will be discussed.

First we note that in the ideal MHD case one cannot integrate over any singularities in the mode spectrum for the three fundamental wave resonances without introducing a complex integration path. One needs to exercise caution as the singularities can be passed over in non-adaptive codes due to numerical inaccuracies. There are two types of modes: internal and external MHD modes. Internal modes exist inside the plasma column. The boundary conditions for internal modes should vanish at the regular radial point or be small at the singular point. The external modes can only exist in the frequency gap in which there is no continuum singularity for a given frequency. For example the so-called TAE gap exists roughly in the range $0.4 < (\omega/\omega_A)^2 < 0.64$ in figure 2.1.

Here, we note that there can actually be no open frequency gap in general. The existence of the TAE gap in figure 2.1 is simply due to the fact that the density profile effect has not been taken into account. Taking into account the dramatic density reduction at the plasma edge, one can find that the Alfvén frequency becomes very large and the gap is usually closed near the plasma edge.

Only searching for the modes in the open gap is insufficient, as internal modes can exist. For example in figure 2.1 one has to determine whether there are internal modes around the $\omega^2/\omega_A^2 = 0.8$ frequency domain in the region $0 < \psi \lesssim 90$. Even in the open gap one has to check the internal instability before determining the external modes. This is similar to the usual procedure for determining MHD stability: one checks the Mercier stability before studying the external kink modes. The capacity to naturally determine both the internal and external global MHD modes for an arbitrary mode frequency is the advantage of the AEGIS code. Its internal mode capacity is in fact an extension to the DCON code [11] for MHD marginal stability.

In the case of continuum singularities, one can use the complex Landau integration path in the configuration space to bypass them. Alternatively, note that, if an analytical function is given on a curve on the complex ω plane, the function can be fully determined in the whole complex domain through the analytical continuation by using the Cauchy–Riemann condition. Note also that one can avoid the MHD continuum by scanning the dispersion relation with real frequency $\Re e\{\omega\}$ for a given small positive growth rate $\Im m\{\omega\}$. Using the scan one can, in principle, find damping roots through the analytical continuation of the dispersion relation. Because of its adaptive shooting scheme the AEGIS code can be used to compute MHD modes with very small growth rates. It was reported in [17] that the Alfvén continuum damping rate was computed using this analytical continuation technique based on the AEGIS code. This technique is also used to compute the marginal stability of low frequency MHD modes in [18]. In fact, for zero frequency MHD modes the rational surfaces, on which the safety factor is a rational number, are always the singular surfaces. The analytical continuation from the small growth rate case can simplify the calculation of external modes, such as the resistive wall modes.

Next, we describe the solution of the plasma region equation (2.20). This solution is then used to determine the internal MHD modes or matched to the vacuum–wall solutions for determining the external MHD modes. Introducing the expanded $2M$ unknowns $\mathbf{u} = \binom{\xi}{\mathbf{u}_2}$, where $\mathbf{u}_2 = \mathcal{F}\boldsymbol{\xi}' + \mathcal{K}\boldsymbol{\xi}$, (2.20) is reduced to the set of $2M$ first order equations

$$\mathbf{u}' = \mathcal{L}\mathbf{u}, \tag{2.24}$$

where the $2M \times 2M$ matrix is

$$\mathcal{L} = \begin{pmatrix} -\mathcal{F}^{-1}\mathcal{K} & \mathcal{F}^{-1} \\ \mathcal{G} - \mathcal{K}^{\mathrm{L}}\mathcal{F}^{-1}\mathcal{K} & \mathcal{K}^{\mathrm{L}}\mathcal{F}^{-1} \end{pmatrix}.$$

One can prove that $\boldsymbol{\xi}$ and $\mathbf{u}_2$ in the plasma region are related to the magnetic field and pressure as follows

$$[[\mathcal{J}\nabla\psi \cdot \delta\mathbf{B}]] = \mathrm{i}Q\boldsymbol{\xi},$$
$$-[[\mathcal{J}(\mathbf{B} \cdot \delta\mathbf{B} - \boldsymbol{\xi} \cdot \nabla P)]] = \mathbf{u}_2.$$

Before describing the solution of (2.20) one needs to determine the boundary conditions. There are three possible types of boundaries at the left end ψ_1, depending on the mode extension. If the modes extend to the magnetic axis we use the cylinder limit to describe the boundary condition at the magnetic axis, i.e. $\xi_{\psi,m} \propto r^m$. If the modes end at the left at a singular surface, the small solution has been chosen. If the modes end at the left at a regular surface, the condition $\boldsymbol{\xi} = 0$ is used.

The set of eigenmode equations in (2.24) can be solved numerically using the independent solution method together with the multiple region matching technique as described in [12]. With M boundary conditions imposed at the left end ψ_1, there remain only M independent solutions:

$$\begin{pmatrix} \Xi_\mathrm{p} \\ \mathcal{W}_2 \end{pmatrix} \equiv \begin{pmatrix} \boldsymbol{\xi}^1, & \cdots, & \boldsymbol{\xi}^M \\ \mathbf{u}_2^1, & \cdots, & \mathbf{u}_2^M \end{pmatrix},$$

where the superscripts are used to label the independent solutions.

The general solution can then be obtained as a combination of the M independent solutions,

$$\begin{pmatrix} \boldsymbol{\xi} \\ \mathbf{u}_2 \end{pmatrix} = \mathrm{i}\begin{pmatrix} \Xi_\mathrm{p} \\ \mathcal{W}_\mathrm{p} \end{pmatrix}\mathbf{c}_\mathrm{p}, \tag{2.25}$$

where $\mathbf{c}_\mathrm{p}$ is a constant vector with M elements. Without loss of generality (by defining $\mathbf{c}_\mathrm{p} = \Xi_\mathrm{p}^{-1}\mathbf{c}_\mathrm{p}^{\mathrm{new}}$ and $\mathcal{W}_\mathrm{p}^{\mathrm{new}} = \mathcal{W}_\mathrm{p}\Xi_\mathrm{p}^{-1}$), we can set Ξ_p to be unity $\mathcal{I}$ or set a certain linear combination of Ξ_p and Ξ_p' to vanish at the right end of the plasma edge ψ_a in the external mode case. Therefore, at the right end we have

$$[[\mathcal{J}\nabla\psi \cdot \delta\mathbf{B}]] = -Q\mathbf{c}_\mathrm{p}, \tag{2.26}$$

$$-[[\mathcal{J}(\mathbf{B} \cdot \delta\mathbf{B} - \boldsymbol{\xi} \cdot \nabla P)]] = \mathrm{i}\mathcal{W}_\mathrm{p}\mathbf{c}_\mathrm{p}. \tag{2.27}$$

The internal modes can be determined if there is a non-trivial solution which satisfies the boundary condition $\boldsymbol{\xi} = 0$ at the right end ψ_r inside the plasma (or $\boldsymbol{\xi}$ is small if the right end is a singular surface $\det|\mathcal{F}| = 0$). Numerically, one can simply scan the value of the determinant $\det|\Xi_\mathrm{p}|$ from the left end to the right end. If $\det|\Xi_\mathrm{p}| = 0$ at a certain ψ_r inside the plasma region for a eigenfrequency, there is an internal mode. The corresponding internal eigenmode can be determined by solving the homogeneous linear equations $\Xi_\mathrm{p}\mathbf{c}_\mathrm{p} = 0$ (or is small if it is a singular surface) at ψ_r, with one of the M components $\mathbf{c}_\mathrm{p}$ chosen as the normalization constant. The eigenfunction is then given by $\boldsymbol{\xi} = \Xi_\mathrm{p}\mathbf{c}_\mathrm{p}$ for $\psi_1 \leqslant \psi \leqslant \psi_\mathrm{r}$. This is a similar procedure to the DCON code for determining the MHD marginal stability [11].

Note that the energy flux at a particular surface ψ_r is proportional to $\boldsymbol{\xi}^\dagger \cdot \mathbf{u}_2|_{\psi_r}$ [12]. When $\boldsymbol{\xi}|_{\psi_r} = 0$, there is no energy flux across the surface ψ_r. Note that the condition $\boldsymbol{\xi} = 0$ implies $\delta B_\psi = 0$. Due to $\nabla \cdot \delta\mathbf{B} = 0$, one can expect that δB_ψ remains vanishing across a thin layer from ψ_r^- to ψ_r^+. In this thin layer a sheet current can be excited to offset the tangential component of magnetic filed, δB_t, on the inner side of the sheet ψ_r^-, so that δB_t vanishes as well on the outer side of the sheet ψ_r^+. Because both δB_ψ and δB_t vanish at ψ_r^+, one can expect that the mode is localized inside the surface ψ_r^+ and outside the sheet plasma is not perturbed. When ψ_r is a singular surface, where det $|\mathcal{F}| = 0$, the conditions $\boldsymbol{\xi} = 0$ and det $|\mathcal{F}| = 0$ yield $\mathbf{u}_2 = 0$ at ψ_r as well. No sheet current is excited in this case. When $\boldsymbol{\xi}$ is small at a singular surface, one can prove that there is no energy flux across the singular surface either. The sheet current, however, may be excited in this case.

Next, we consider the external modes, in particular the resistive wall modes. For these one requires the vacuum solutions [12]. The vacuum regions are described by the Laplace equation

$$\nabla^2 u = 0, \tag{2.28}$$

where u is the magnetic scalar potential and is related to the perturbed magnetic field by $\delta\mathbf{B} = -\nabla u$. Here, we note that this representation of the vacuum magnetic field, although simple, excludes the consideration of $n = 0$ modes. To study the $n = 0$ modes, an additional scalar is needed to represent the vacuum magnetic field. This can be found in [19]. Here, we limit ourselves to the $n \neq 0$ case. For the sake of conciseness, we outline the general solutions for the inner and outer vacuum regions simultaneously.

As in the plasma region, the Fourier decompositions are introduced for both the poloidal and toroidal directions to solve (2.28). Then (2.28) becomes a set of second order differential equations of number M for $\mathbf{u}$. This set of second order differential equations can be transformed into a set of first order differential equations of number $2M$ by introducing an additional field $\mathbf{v} = -[[\mathcal{J}\nabla\psi \cdot \delta\mathbf{B}]]$, which is related to the magnetic scalar potential in the Fourier space as follows:

$$\mathbf{v} = \left\langle \mathcal{J}\,|\nabla\psi|^2 \right\rangle \frac{\partial \mathbf{u}}{\partial \psi} + \langle \mathrm{i}\mathcal{J}\nabla\psi \cdot \nabla\theta \rangle \mathcal{M}\mathbf{u}.$$

There are $2M$ independent solutions for (2.28), which can be used to construct the following independent solution matrices:

$$\begin{pmatrix} \mathcal{U}_1 \\ \mathcal{V}_1 \end{pmatrix} \equiv \begin{pmatrix} \mathbf{u}^1, & \cdots, & \mathbf{u}^M \\ \mathbf{v}^1, & \cdots, & \mathbf{v}^M \end{pmatrix},$$

$$\begin{pmatrix} \mathcal{U}_2 \\ \mathcal{V}_2 \end{pmatrix} \equiv \begin{pmatrix} \mathbf{u}^{M+1}, & \cdots, & \mathbf{u}^{2M} \\ \mathbf{v}^{M+1}, & \cdots, & \mathbf{v}^{2M} \end{pmatrix}.$$

These are distinguished by the applied boundary conditions (to be specified below). The general solutions in the vacuum regions can be expressed as a linear combination of the independent solutions:

$$\begin{pmatrix} \mathbf{u} \\ \mathbf{v} \end{pmatrix} = \begin{pmatrix} \mathcal{U}_1 \\ \mathcal{V}_1 \end{pmatrix} \mathbf{c}_{\mathrm{v}} + \begin{pmatrix} \mathcal{U}_2 \\ \mathcal{V}_2 \end{pmatrix} \mathbf{d}_{\mathrm{v}}, \tag{2.29}$$

where $\mathbf{c}_{\mathrm{v}}$ and $\mathbf{d}_{\mathrm{v}}$ are constant vectors in the independent solution space. To distinguish the inner and outer vacuum solutions, we use $\mathbf{c}_{\mathrm{v}1}$ and $\mathbf{d}_{\mathrm{v}1}$ to denote the constants for the inner vacuum region and $\mathbf{c}_{\mathrm{v}2}$ and $\mathbf{d}_{\mathrm{v}2}$ for the outer vacuum region.

In the outer vacuum region, the scalar potential $\mathbf{u}$ is subjected to M boundary conditions at infinite ψ: $\mathbf{u} \rightarrow 0$ (or is the small solution). With these M boundary conditions imposed, there are only M independent solutions left. Without loss of generality, we can set $\mathbf{c}_{\mathrm{v}2}$ to be zero in this case. Consequently, eliminating $\mathbf{d}_{\mathrm{v}2}$ in (2.29), we obtain

$$\mathbf{u}|_{\psi b+} = \mathcal{T}\mathbf{v}|_{\psi b+},$$

where the $M \times M$ matrix $\mathcal{T}$ is given by $\mathcal{T} = \mathcal{U}_2 \mathcal{V}_2^{-1}|_{\psi b+}$. The matrix $\mathcal{T}$ can be computed by means of the Green's function method [20].

In the inner vacuum region, independent solutions can be constructed, for example, using inward numerical shooting [12], with the following boundary conditions imposed at ψ_{b-}:

$$\begin{pmatrix} \mathcal{U}_1 \\ \mathcal{V}_1 \end{pmatrix}_{\psi b-} = \begin{pmatrix} \mathcal{I} \\ \mathcal{O} \end{pmatrix}, \tag{2.30}$$

$$\begin{pmatrix} \mathcal{U}_2 \\ \mathcal{V}_2 \end{pmatrix}_{\psi b-} = \begin{pmatrix} \mathcal{T} \\ \mathcal{I} \end{pmatrix}, \tag{2.31}$$

where $\mathcal{O}$ is an $M \times M$ zero matrix. Since the boundary conditions in (2.30) give $\delta\mathbf{B} \cdot \nabla\psi = 0$ at the wall, these conditions correspond to the set of solutions that corresponds to the perfectly conducting wall type. On the other hand, since the boundary conditions in (2.31) guarantee that the independent solutions are continuous with the outer vacuum solutions, these conditions correspond to the set of solutions that corresponds to the no-wall type. Using the general expression for the solutions in (2.29), we can express the normal and parallel magnetic fields at the plasma–vacuum interface as follows:

$$[[\mathcal{J}\nabla\psi \cdot \delta\mathbf{B}]] = -\mathcal{V}_1\mathbf{c}_{\mathrm{v}1} - \mathcal{V}_2\mathbf{d}_{\mathrm{v}1}, \tag{2.32}$$

$$-[[\mathcal{J}\mathbf{B} \cdot \delta\mathbf{B}]] = \mathrm{i}\mathcal{Q}\left(\mathcal{U}_1\mathbf{c}_{\mathrm{v}1} + \mathcal{U}_2\mathbf{d}_{\mathrm{v}1}\right). \tag{2.33}$$

The solutions in the plasma and vacuum regions can be used to construct the eigenvalue problem [12]. The normal magnetic field component and the combined magnetic and thermal pressures are required to be continuous at the plasma–vacuum interface. Matching the plasma ((2.26) and (2.27)) and vacuum ((2.32) and (2.33)) solutions at the interface ψ_a gives

$$\mathbf{d}_{v1} = \mathcal{F}_1^{-1}\delta\mathcal{W}_b\delta\mathcal{W}_\infty^{-1}\mathcal{F}_2\mathbf{c}_{v1}, \tag{2.34}$$

where $\delta\mathcal{W}_\infty = \mathcal{W}_p - Q[\mathcal{U}_2\mathcal{V}_2^{-1}]_{\psi_a}Q$, $\delta\mathcal{W}_b = \mathcal{W}_p - Q[\mathcal{U}_1\mathcal{V}_1^{-1}]_{\psi_a}Q$, $\mathcal{F}_1 = Q[\mathcal{U}_2 - \mathcal{U}_1\mathcal{V}_1^{-1}\mathcal{V}_2]_{\psi_a}$, and $\mathcal{F}_2 = Q[\mathcal{U}_1 - \mathcal{U}_2\mathcal{V}_2^{-1}\mathcal{V}_1]_{\psi_a}$. Note that $\delta\mathcal{W}_\infty$ and $\delta\mathcal{W}_b$ correspond to the energy matrices without a wall and with a perfectly conducting wall at ψ_b, respectively, as can be seen from the boundary conditions in (2.30) and (2.31).

We now consider the matching across the thin resistive wall. The Maxwell equation $\nabla \cdot \delta\mathbf{B} = 0$ and the thin wall assumption yield that the radial magnetic field is continuous across the wall. Using this condition, one obtains

$$\mathbf{v}|_{\psi b-} = \mathbf{v}|_{\psi b+} = \mathbf{d}_{v1}.$$

The current in the resistive wall causes a jump in the scalar magnetic potential. This can be obtained from the Ampére law

$$\nabla \times \nabla \times \delta\mathbf{B} = -\gamma\mu_0\sigma\delta\mathbf{B}, \tag{2.35}$$

where σ is the wall conductivity. Equation (2.35) can be reduced to

$$\mathcal{V}\left(\mathbf{u}|_{\psi b+} - \mathbf{u}|_{\psi b-}\right) = \tau_w\gamma_N\mathbf{d}_{v1}, \tag{2.36}$$

where the resistive wall time $\tau_w = \mu_0\sigma db/\tau_A$, d is the wall thickness, b is the average wall minor radius, and

$$\mathcal{V} = \mathcal{M}\left\langle \mathcal{J}\,|\nabla\psi||\nabla\theta| - \mathcal{J}\,|\nabla\psi\cdot\nabla\theta|^2/(|\nabla\psi||\nabla\theta|)\right\rangle\mathcal{M}$$
$$+\, n^2\left\langle \mathcal{J}\,|\nabla\phi|^2|\nabla\psi|/|\nabla\theta|\right\rangle.$$

Since $\mathbf{c}_{v2} = 0$, we find that (2.29)–(2.31) yield

$$\mathbf{u}\,|_{\psi b+} - \mathbf{u}|_{\psi b-} = -\mathbf{c}_{v1}. \tag{2.37}$$

from (2.34), (2.36) and (2.37) we find the eigenmode equations

$$\mathcal{D}_0(\gamma_N)\mathbf{d}_{v1} \equiv \tau_w\gamma_N\mathbf{d}_{v1} + \mathcal{V}\mathcal{F}_2^{-1}\delta\mathcal{W}_\infty\delta\mathcal{W}_b^{-1}\mathcal{F}_1\mathbf{d}_{v1} = O.$$

The dispersion relation for this eigenvalue problem is given by the determinant equation det $|\mathcal{D}_0(\gamma_N)| = 0$. In general the Nyquist diagram or the analytical continuation method can be used to determine the roots of this dispersion relation. For resistive wall modes, however, the growth rate is much smaller than the Alfvén frequency. Therefore, the growth rate dependence of $\delta\mathcal{W}_\infty\delta\mathcal{W}_b^{-1}$ can be neglected in determining the stability condition. Consequently, one can use the reduced eigenvalue problem

$$-\mathcal{V}\mathcal{F}_2^{-1}\delta\mathcal{W}_\infty\delta\mathcal{W}_b^{-1}\mathcal{F}_1\mathbf{d}_{v1} = \tau_w\gamma_N\mathbf{d}_{v1}, \tag{2.38}$$

with the resistive wall mode growth rate γ_N on the right-hand side of the equation used as the eigenvalue to determine the stability.

2.3 Radially localized modes: the Mercier criterion

In this section we study the interchange mode theory in a toroidal geometry. One of the most fundamental phenomena in magnetically confined plasmas is the interchange mode, which resembles the so-called Rayleigh–Taylor instability in conventional fluid theory. The interchange of plasma flux tubes can lead to the release of plasma thermal energy and thus an instability develops. The perturbed magnetic energy from field line bending is minimized for the interchange instability. In the slab or cylinder configurations such an interchange occurs due to the existence of the so-called bad curvature region. In a toroidal geometry, however, the finite safety factor value q causes the magnetic curvature direction, with respect to the plasma pressure gradient, to alternate between the low and high field sides of the plasma torus. Therefore, a toroidal average becomes necessary in evaluating the change in the plasma and magnetic energies. This makes the interchange mode theory in the plasma torus complicated. This interchange mode theory was one of the first successful toroidal theories in the field. It includes the derivations of the so-called singular layer equation and the interchange stability criterion, i.e. the so-called Mercier criterion [21, 22].

The previous derivation of a singular layer equation relied on the assumption that the modes are somewhat poloidally localized. This assumption was removed in a paper by A Glasser *et al* [23]. However, the details are omitted in their paper and a direct projection method, an alternative to the original vorticity equation approach [25], is used. Here, we detail the derivation of the singular layer equation using the vorticity equation approach as in [1]. The derivation details the analytical techniques used to separate the compressional Alfvén wave from the low frequency interchange mode and to minimize field line bending effects. The singular layer equation will be used to derive the stability criteria for the interchange modes, as well as the peeling modes, in section 2.5.

Before deriving the Mercier criterion, we first introduce some fundamental formulas [24] that will be used in the derivation. Noting that $\nabla(2\mu_0 P + B^2) = 2\mathbf{B}\cdot\nabla\mathbf{B}$, the field line curvature can be expressed as

$$\boldsymbol{\kappa} = \mathbf{b}\cdot\nabla\mathbf{b} = \frac{\left[\mathbf{B}\times\nabla\left(2\mu_0 P + B^2\right)\right]\times\mathbf{B}}{2B^4}. \tag{2.39}$$

We also have

$$\frac{2\mathbf{B}\times\nabla P\cdot\boldsymbol{\kappa}}{B^4} = \frac{\mathbf{B}\times\nabla P\cdot\nabla\left(2\mu_0 P + B^2\right)}{B^4} = \mathbf{B}\cdot\nabla\sigma. \tag{2.40}$$

Here, σ corresponds to the parallel current density as given in (1.11). In addition, we will also use the following three projections of perturbed magnetic field:

$$\delta\mathbf{B}\cdot\nabla Z = \mathbf{B}\cdot\nabla(\boldsymbol{\xi}\cdot\nabla Z), \tag{2.41}$$

$$\frac{\delta\mathbf{B}\cdot\mathbf{B}\times\nabla Z}{|\nabla Z|^2}=\mathbf{B}\cdot\nabla\left(\frac{\boldsymbol{\xi}\cdot\mathbf{B}\times\nabla Z}{|\nabla Z|^2}\right)$$

$$-\frac{(\mathbf{B}\times\nabla Z)\cdot\nabla\times(\mathbf{B}\times\nabla Z)}{|\nabla Z|^4}\boldsymbol{\xi}\cdot\nabla Z, \tag{2.42}$$

$$\frac{1}{B^2}\left(\delta\mathbf{B}-\mathbf{B}\frac{\mu_0\boldsymbol{\xi}\cdot\nabla P}{B^2}\right)\cdot\mathbf{B}=-\nabla\cdot\boldsymbol{\xi}_\perp-2\boldsymbol{\kappa}\cdot\boldsymbol{\xi}. \tag{2.43}$$

Here, Z is an arbitrary label for the magnetic surfaces and $\delta\mathbf{B}=\nabla\times\boldsymbol{\xi}\times\mathbf{B}$ represents the ideal MHD case.

To investigate the modes which localize around a particular rational (or singular) magnetic surface V_0, we specialize the Hamada coordinates to the neighborhood of the mode rational surface V_0 and introduce the localized Hamada coordinates $x,\ u,\ \theta$ as usual, where $x=V-V_0$ and $u=m\theta-n\zeta$. In this coordinate system the parallel derivative becomes $\mathbf{B}\cdot\nabla=\psi'(\partial/\partial\zeta)+(\Lambda x/\Xi)(\partial/\partial u)$, where $\Lambda=\psi'\chi''-\chi'\psi''$ and $\Xi=\psi'/m=\chi'/n$.

Noting that the equilibrium scalars are independent of u in the axisymmetric system, the perturbations can be assumed to vary as $\exp\{i\alpha u\}$ with $\alpha=2\pi n/\chi'$. As in [25] and [23], $\boldsymbol{\xi}$ and $\delta\mathbf{B}$ are projected in three directions as follows:

$$\boldsymbol{\xi}=\xi\frac{\nabla V}{|\nabla V|^2}+\mu\frac{\mathbf{B}\times\nabla V}{B^2}+\nu\frac{\mathbf{B}}{B^2},$$

$$\delta\mathbf{B}=b\frac{\nabla V}{|\nabla V|^2}+v\frac{\mathbf{B}\times\nabla V}{B^2}+\tau\frac{\mathbf{B}}{B^2}.$$

We consider only the low-frequency regime

$$|\omega/\omega_{si}|\ll 1. \tag{2.44}$$

where ω_{si} is the parallel ion acoustic frequency. It is assumed that the mode wavelength across the magnetic surface $\lambda_\perp$ is much smaller than that on the surface and perpendicular to the magnetic field line $\lambda_\wedge$. This leads us to introduce the following ordering scheme as in [23]:

$$x\sim\varepsilon,\qquad \frac{\partial}{\partial V}\sim\varepsilon^{-1},\qquad \frac{\partial}{\partial u}\sim\frac{\partial}{\partial\theta}\sim 1,$$

where $\epsilon\ll 1$ is a small parameter.

Since the modes vary on a slow time scale, they are decoupled from the compressional Alfvén wave. It can be verified *a posteriori* that we can make the following ordering assumptions:

$$\xi=\epsilon\xi^{(1)}+\cdots,\quad \mu=\mu^{(0)}+\cdots,\quad \delta\hat{P}^{(2)}=\epsilon^2\delta\hat{P}^{(2)}+\cdots,$$
$$b=\epsilon^2 b^{(2)}+\cdots,\quad v=\epsilon v^{(1)}+\cdots,\quad \tau=\epsilon\tau^{(1)}+\cdots,$$

where $\delta \hat{P}^{(2)} = -\Gamma P \nabla \cdot \boldsymbol{\xi}$. These ordering assumptions are the same as those in [23], except we use $\delta \hat{P}^{(2)}$ as an unknown to replace ν. As discussed below, the ordering assumption for $\delta \hat{P}^{(2)}$ is consistent with the low frequency assumption in (2.44). With these ordering assumptions we can proceed to analyze the basic set of linearized MHD equations (2.1)–(2.4). As usual, the perturbed quantities are separated into constant and oscillatory parts along the field lines: $\xi = \bar{\xi} + \tilde{\xi}$, where $\bar{\xi} = \langle \xi \rangle \equiv \oint \xi dl/B / \oint dl/B$, l is the arc length of the magnetic field line and $\tilde{\xi} = \xi - \langle \xi \rangle$.

The condition that $\delta \mathbf{B}$ is divergence-free, (2.2), yields

$$\frac{\partial b^{(2)}}{\partial x} + \frac{1}{\Xi}\frac{\partial}{\partial u}v^{(1)} + \frac{\partial v^{(1)}}{\partial \theta}\frac{\mathbf{B} \times \nabla V \cdot \nabla \theta}{B^2} + v^{(1)} \nabla \cdot \frac{\mathbf{B} \times \nabla V}{B^2} + \chi' \frac{\partial}{\partial \theta}\frac{\tau^{(1)}}{B^2} = 0. \quad (2.45)$$

Note that

$$\frac{\mathbf{B} \times \nabla V \cdot \nabla \theta}{B^2} = \frac{1}{P'}(\mathbf{J} - \sigma \mathbf{B}) \cdot \nabla \theta = \frac{J'}{P'} - \frac{\chi'}{P'}\sigma,$$

$$\nabla \cdot \frac{\mathbf{B} \times \nabla V}{B^2} = \frac{2\mathbf{B} \times \boldsymbol{\kappa}}{B^2} \cdot \nabla V = -\frac{1}{P'}\chi'\frac{\partial}{\partial \theta}\sigma, \quad (2.46)$$

where (2.40) has been used. Equation (2.45) can be reduced to

$$\frac{\partial b^{(2)}}{\partial x} + \frac{1}{\Xi}\frac{\partial}{\partial u}v^{(1)} + \frac{J'}{P'}\frac{\partial v^{(1)}}{\partial \theta} - \frac{\chi'}{P'}\frac{\partial \sigma v^{(1)}}{\partial \theta} + \chi'\frac{\partial}{\partial \theta}\frac{\tau^{(1)}}{B^2} = 0. \quad (2.47)$$

After the surface average it gives

$$\frac{\partial \bar{b}^{(2)}}{\partial x} + \frac{1}{\Xi}\frac{\partial \bar{v}^{(1)}}{\partial u} = 0. \quad (2.48)$$

The two significant orders of the induction equation, (2.2), in the ∇V-direction are

$$0 = \chi'\frac{\partial \xi^{(1)}}{\partial \theta}, \quad (2.49)$$

$$b^{(2)} = \chi'\frac{\partial \xi^{(2)}}{\partial \theta} + \frac{\Lambda x}{\Xi}\frac{\partial \xi^{(1)}}{\partial u}, \quad (2.50)$$

where (2.41) has been used. The component of (2.2) in the ∇u-direction, in the lowest order, yields

$$\chi'\frac{\partial \mu^{(0)}}{\partial \theta} = 0. \quad (2.51)$$

To satisfy the component of (2.2) which is along the magnetic field line, one must require that

$$(\nabla \cdot \boldsymbol{\xi}_\perp)^{(0)} + 2\boldsymbol{\kappa} \cdot \boldsymbol{\xi}^{(0)} = \frac{\partial \xi^{(1)}}{\partial x} + \frac{1}{\Xi}\frac{\partial \mu^{(0)}}{\partial u} = 0, \quad (2.52)$$

where (2.43), (2.46) and (2.51) have been used.

Next, we turn to the momentum equation (2.5). The two components perpendicular to the $\mathbf{B}$ of the momentum equation (2.5) both lead, in the lowest order, to

$$\tau^{(1)} - P'\xi^{(1)} = 0. \tag{2.53}$$

This is consistent with (2.52). Noting that the two perpendicular components of (2.5) yield the same information, we can directly work on the vorticity equation (2.9) and obtain in the leading and first orders

$$\chi'\frac{\partial}{\partial\theta}\left(\frac{|\nabla V|^2}{B^2}\frac{\partial v^{(1)}}{\partial x}\right) + \chi'\frac{\partial\sigma}{\partial\theta}\frac{\partial\xi^{(1)}}{\partial x} = 0, \tag{2.54}$$

$$\begin{aligned}
&-\omega^2\frac{\rho_m}{B^2}\frac{|\nabla V|^2}{\partial x}\frac{\partial\mu^{(0)}}{\partial x}\\
&= -\chi'\frac{\partial}{\partial\theta}\left(\frac{|\nabla V|^2}{B^2}\frac{\partial v^{(2)}}{\partial x} - v\frac{\mathbf{B}}{B^2}\cdot\nabla\times\frac{\mathbf{B}\times\nabla V}{B^2} - \tau\frac{\mathbf{B}}{B^2}\cdot\nabla\times\frac{\mathbf{B}}{B^2} + \frac{J'}{\chi'}\tau^{(1)}\right)\\
&\quad - v^{(1)}\left(\frac{J'}{P'} - \frac{\chi'}{P'}\sigma\right)\frac{\partial\sigma}{\partial\theta} - \tau^{(1)}\frac{\chi'}{B^2}\frac{\partial\sigma}{\partial\theta} - \Lambda x\frac{|\nabla V|^2}{\Xi B^2}\frac{\partial}{\partial u}\frac{\partial v^{(1)}}{\partial x} + \frac{P'}{\Xi B^2}\frac{\partial\tau^{(1)}}{\partial u}\\
&\quad + P'\frac{\nabla V\cdot\nabla\left(P + B^2\right)}{\Xi B^2|\nabla V|^2}\frac{\partial\xi^{(1)}}{\partial u} - \chi'\frac{\partial\sigma}{\partial\theta}\Theta\frac{\partial\xi^{(1)}}{\partial u} + \frac{\chi'}{P'}\frac{\partial\sigma}{\partial\theta}\frac{\partial}{\partial x}\left(\delta\hat{P}^{(2)} - P'\xi^{(2)}\right).
\end{aligned} \tag{2.55}$$

We will derive the singular layer equation by averaging (2.55). This needs to express all unknowns in this equation in terms of $\xi^{(1)}$.

It is trivial to obtain $\mu^{(0)}$ from (2.51) and (2.52), and $\tau^{(1)}$ from (2.53). The rest can be obtained as follows. From (2.49) and (2.50) one can find that $\bar{b}^{(2)} = (\Lambda x/\Xi)(\partial\xi^{(1)}/\partial u)$. With $\bar{b}^{(2)}$ obtained one can determine $\bar{v}^{(1)}$ from (2.48):

$$\bar{v}^{(1)} = -\Lambda\frac{\partial}{\partial x}\left(x\xi^{(1)}\right). \tag{2.56}$$

Using (2.56) to determine the integration constant, (2.54) can be solved, yielding that

$$\frac{\partial v^{(1)}}{\partial x} = -\left(\frac{B^2\sigma}{|\nabla V|^2} - \frac{\left\langle B^2\sigma/|\nabla V|^2\right\rangle}{\left\langle B^2/|\nabla V|^2\right\rangle}\frac{B^2}{|\nabla V|^2}\right)\frac{\partial\xi^{(1)}}{\partial x} - \Lambda\frac{B^2/|\nabla V|^2}{\left\langle B^2/|\nabla V|^2\right\rangle}\frac{\partial^2}{\partial x^2}\left(x\xi^{(1)}\right). \tag{2.57}$$

From (2.50) and (2.47) one obtains

$$-\chi'\frac{\partial^2\xi^{(2)}}{\partial\theta\partial x} = \frac{1}{\Xi}\frac{\partial\tilde{v}^{(1)}}{\partial u} + \frac{J'}{P'}\frac{\partial v^{(1)}}{\partial\theta} - \frac{\chi'}{P'}\frac{\partial\sigma v^{(1)}}{\partial\theta} + \chi'\frac{\partial}{\partial\theta}\frac{\tau^{(1)}}{B^2}. \tag{2.58}$$

We also need to solve the equation of parallel motion, (2.8). Taking into consideration the low frequency assumption in (2.44) and the result in (2.52), the equation of parallel motion can be reduced to

$$\chi'^2 \frac{\partial}{\partial\theta}\left(\frac{1}{B^2}\frac{\partial}{\partial\theta}\delta\hat{P}^{(2)}\right) = \mathrm{i}\frac{\rho_m\omega^2}{\alpha\Gamma P}\frac{\mathbf{B}\times\nabla V}{B^2}\cdot\boldsymbol{\kappa}\frac{\partial\xi^{(1)}}{\partial x}. \tag{2.59}$$

Using (2.46), equation (2.59) can be solved to yield

$$\chi'\frac{\partial}{\partial\theta}\delta\hat{P}^{(2)} = \mathrm{i}\frac{\rho_m\omega^2}{\alpha\Gamma P}\left(B^2\sigma - \frac{\left\langle B^2\sigma\right\rangle}{\left\langle B^2\right\rangle}B^2\right)\frac{\partial\xi^{(1)}}{\partial x}.$$

This indicates that the oscillating part of $\delta\hat{P}^{(2)}$ is of order ϵ^2. To prove that the average part of $\delta\hat{P}^{(2)}$ also is of order ϵ^2 from the equation of parallel motion (2.8), one needs to use the low frequency assumption in (2.44) as an auxiliary ordering.

Furthermore, to reduce the term $\langle\chi'\frac{\partial\sigma}{\partial\theta}\Theta\frac{\partial\xi^{(1)}}{\partial u}\rangle$ in (2.55), we note that

$$-\sigma\chi'\frac{\partial}{\partial\theta}\frac{\nabla V\cdot\nabla u}{|\nabla V|^2} = -\sigma(\psi'\chi'' - \chi'\psi'') + \frac{\sigma\mathbf{B}\times\nabla V\cdot\nabla\times(\mathbf{B}\times\nabla V)}{|\nabla V|^4}$$

and [25]

$$\begin{aligned}
&\frac{\mathbf{B}\times\nabla V\cdot\nabla\times(\mathbf{B}\times\nabla V)}{|\nabla V|^4}\\
&= \frac{\mathbf{B}\times\nabla V}{|\nabla V|^2}\cdot\nabla\times\left(\frac{\nabla V\times(\psi'\nabla\theta - \chi'\nabla\zeta)]\times\nabla V}{|\nabla V|^2}\right),\\
&= \frac{(\mathbf{B}\times\nabla V)\times\nabla V}{|\nabla V|^2}\cdot\left[\psi''\nabla\theta - \chi''\nabla\zeta + \nabla\left(\frac{\nabla\psi\cdot\nabla\theta - \nabla\chi\cdot\nabla\zeta}{|\nabla V|^2}\right)\right]\\
&= \psi'\chi'' - \chi'\psi'' - \mathbf{B}\cdot\nabla\left(\frac{\nabla\psi\cdot\nabla\theta - \nabla\chi\cdot\nabla\zeta}{|\nabla V|^2}\right).
\end{aligned} \tag{2.60}$$

Finally, the average over l of (2.55) yields the singular layer equation

$$\frac{\partial}{\partial x}\left(x^2 - M\omega^2\right)\frac{\partial\xi^{(1)}}{\partial x} + \left(\frac{1}{4} + D_{\mathrm{I}}\right)\xi^{(1)} = 0, \tag{2.61}$$

where the total mass parameter $M = M_{\mathrm{c}} + M_{\mathrm{t}}$,

$$D_{\mathrm{I}} \equiv E + F + H - \frac{1}{4}, \tag{2.62}$$

$$E \equiv \frac{\left\langle B^2/|\nabla V|^2\right\rangle}{\Lambda^2}\left(J'\psi'' - I'\chi'' + \Lambda\frac{\left\langle \sigma B^2\right\rangle}{\left\langle B^2\right\rangle}\right),$$

$$F \equiv \frac{\left\langle B^2/|\nabla V|^2\right\rangle}{\Lambda^2}\left(\left\langle \frac{\sigma^2 B^2}{|\nabla V|^2}\right\rangle - \frac{\left\langle \sigma B^2/|\nabla V|^2\right\rangle^2}{\left\langle B^2/|\nabla V|^2\right\rangle} + P'^2\left\langle \frac{1}{B^2}\right\rangle\right), \qquad (2.63)$$

$$H \equiv \frac{\left\langle B^2/|\nabla V|^2\right\rangle}{\Lambda}\left(\frac{\left\langle \sigma B^2/|\nabla V|^2\right\rangle}{\left\langle B^2/|\nabla V|^2\right\rangle} - \frac{\left\langle \sigma B^2\right\rangle}{\left\langle B^2\right\rangle}\right),$$

$$M_{\mathrm{c}} \equiv \frac{\rho_m}{\alpha^2\Lambda^2}\left\langle \frac{B^2}{|\nabla V|^2}\right\rangle\left\langle \frac{|\nabla V|^2}{B^2}\right\rangle,$$

$$M_{\mathrm{t}} \equiv \frac{\rho_m}{\alpha^2\Lambda^2 P'^2}\left\langle \frac{B^2}{|\nabla V|^2}\right\rangle\left(\left\langle \sigma^2 B^2\right\rangle - \frac{\left\langle \sigma B^2\right\rangle^2}{\left\langle B^2\right\rangle}\right). \qquad (2.64)$$

The Mercier index in (2.62) is expressed in Hamada coordinates. Alternatively, it can be expressed as

$$D_{\mathrm{I}} = \frac{\langle g\rangle}{\left\langle \mathbf{B}\cdot\nabla\Lambda_s\right\rangle^2}\left[\left\langle P'\kappa_n\right\rangle + \left\langle \mathbf{B}\cdot\nabla\Lambda_s\left(\lambda_{\mathrm{c}} - \frac{\langle g\lambda_{\mathrm{c}}\rangle}{\langle g\rangle}\right)\right\rangle + \left\langle g\left(\lambda_{\mathrm{c}}^2 - \frac{\langle g\lambda_{\mathrm{c}}\rangle^2}{\langle g\rangle^2}\right)\right\rangle\right] - \frac{1}{4}, \qquad (2.65)$$

where

$$g = \frac{B^2}{|\nabla\psi|^2\sigma}, \quad \kappa_n = \frac{\nabla\psi\cdot\boldsymbol{\kappa}}{|\nabla\psi|^2}, \quad \kappa_g = \frac{\mathbf{B}\times\nabla\psi\cdot\boldsymbol{\kappa}}{B^2}, \quad \lambda_{\mathrm{c}} = P'\int_{-\pi}^{l}\kappa_g\frac{\mathrm{d}l}{B},$$

$$\mathbf{B}\cdot\nabla\Lambda_s = -\frac{1}{|\nabla\psi|^4}(\mathbf{B}\times\nabla\psi)\cdot\nabla\times(\mathbf{B}\times\nabla\psi).$$

The expression in (2.65) is often obtained from the asymptotic analysis of the ballooning mode equation [26, 27].

Note that the mass factor M_{c} in (2.63) results from the perpendicular motion and M_{t} from the parallel motion due to the toroidal coupling. M_{t} in (2.64) is often referred to as the apparent mass. In the kinetic description the apparent mass is enhanced by the so-called small parallel ion speed effect. In large aspect ratio configurations this enhancement factor is of order $\sqrt{R/a}$, where R and a are, respectively, the major and minor radii [2, 28].

From (2.61) one can derive the Mercier criterion, i.e. the stability criterion for localized interchange modes in the toroidal geometry. In the marginal stability $\omega^2 = 0$, (2.61) becomes the Euler differential equation. Its solution is

$$\xi = \xi_0 x^{-\frac{1}{2} \pm \sqrt{-D_{\mathrm{I}}}}. \tag{2.66}$$

The system stability can be determined by Newcomb's theorem 5 [29]: the system is unstable, if and only if the solution of (2.61) vanishes at two or more regular points or is small if they are singular points. From the solution in (2.66) one can see that, when $-D_{\mathrm{I}} < 0$, ξ becomes oscillating. Therefore, the interchange mode stability criterion, or Mercier's criterion, is simply $-D_{\mathrm{I}} > 0$.

It is interesting to examine the Mercier criterion in the tokamak configuration with a circular cross section and a large aspect ratio, using the equilibrium described in section 1.3. To simplify the notation, we introduce the dimensionless parameters: $s = \dot{q} r'/q$, $\alpha_{\mathrm{c}} = -2R_0 r'^2 \mu_0 \dot{P}/\psi^2$, and $\psi = RB_\theta$, with the dot denoting the derivative with respect to r'. Hence, the equilibrium equation (1.35) can be reduced to

$$r'\ddot{\xi}_1 + 3\dot{\xi}_1 - 2s\dot{\xi}_1 + \frac{r'}{R_0} + \alpha_{\mathrm{c}} - \frac{\mu_0 r'}{B_\theta^2} j_\phi B_z = 0. \tag{2.67}$$

To evaluate the Mercier criterion, one needs to express the related functions in the criterion (2.65) in terms of the magnetic surface shift ξ_1. First, we note that the arc length along the magnetic field line can be expressed as [30]

$$\frac{\mathrm{d}l}{B} = -\frac{R_0 r'}{\psi}\left(1 + \dot{\xi}_1 \cos\theta + \frac{\xi_1}{r'}\cos\theta + \frac{r'}{R_0}\cos\theta\right)\mathrm{d}\theta.$$

For normal and geodesic curvatures, by noting that $\boldsymbol{\kappa} \approx -\mathbf{R}/R^2 - \left(r'/q^2 R_0^2\right)\nabla r'$, one obtains

$$\kappa_n = \frac{1}{R_0 \psi}\left(-\cos\theta + \frac{r'}{R_0}\cos^2\theta - \dot{\xi}_1 \cos\theta\cos\theta \right.$$
$$\left. + \frac{\xi_1}{r'}\sin\theta\sin\theta - \frac{r'}{R_0 q^2}\right),$$

$$\kappa_g = -\frac{r'}{qR_0}\left(\sin\theta - \frac{r'}{R_0}\cos\theta\sin\theta - \dot{\xi}_1\cos\theta\sin\theta \right.$$
$$\left. + \frac{1}{r'}\xi_1 \sin\theta\cos\theta\right).$$

After some laborious calculations, one can express the shear parameter as

$$\mathbf{B} \cdot \nabla \Lambda_s = \frac{1}{R_0 r'}\left[\dot{q} - 2\dot{q}\frac{r'}{R_0}\cos\theta + q\left(\ddot{\xi}_1 \cos\theta - \frac{1}{R_0}\cos\theta + \frac{\dot{\xi}_1}{r'}\cos\theta\right)\right].$$

Other related functions can be evaluated as well, yielding

$$g = \frac{q^2}{r'^2}\left(1 - \frac{2r'}{R_0}\cos\theta + 2\dot{\xi}_1\cos\theta\right),$$

$$\lambda_c = \dot{P}\frac{r'^2}{q\dot{\psi}}(\cos\theta + 1).$$

Using these results, one can obtain

$$D_I = \frac{\alpha_c}{s^2}\left(\frac{r'}{R_0 q^2} + \frac{1}{2}r'\ddot{\xi}_1 + \frac{3}{2}\dot{\xi}_1 - s\dot{\xi}_1 - \frac{1}{2}\frac{r'}{R_0} + \frac{\alpha_c}{2}\right) - \frac{1}{4}.$$

Here, we consider only the region with finite r, where $B_\theta^2 \gg B_z^2$. By making use of the equilibrium equation (2.67), this formula can be simplified to

$$-D_I = \frac{1}{4} - \frac{1}{s^2}\frac{r'}{R_0}\left(\frac{1}{q^2} - 1 + \frac{\mu_0 R}{2B_\theta^2} j_\phi B_z\right)\alpha_c, \tag{2.68}$$

where the first two terms in the curved brackets on the right-hand side is the well-known ideal MHD correspondent [31, 32] and the final term is due to the additional magnetic surface shift contributed by the external magnetic field B_z. Equation (2.68) indicates that a suitable application of the external vertical magnetic field, which counterbalances the hoop force, can give rise to a stabilizing effect on the interchange modes in the tokamak. This effect can also have positive impacts on other internal modes, such as the kink and ballooning modes.

2.4 The ballooning modes and ballooning representation

In this section, we describe the high-n ballooning mode theory [26, 33]. Ballooning mode theory is one of the most beautiful theories in tokamak physics. Its beauty lies in both its physical and mathematical aspects.

In terms of physics, it is well known that tokamak plasmas have good and bad curvature regions, referring to whether the pressure gradient and magnetic field line curvature point in the same direction or not. The stability criterion for interchange modes takes into account only the average magnetic well effect. Although tokamaks are usually designed to have average good curvature, i.e. be Mercier stable, ballooning modes can still develop as soon as the release of plasma thermal energy in the bad curvature region is sufficient to counter the magnetic energy resulting from the field line bending effects.

In terms of mathematics, the invention of the ballooning mode representation represents a mathematical breakthrough in dealing with the high-n modes. The discovery of ballooning invariance and its application to reduce the two-dimensional problem into a one-dimensional problem shows that, even in a field in which one often has to cope with lengthy or sometimes 'ugly' modeling, simplicity and beauty can still exist. Unlike the interchange mode theory, the ballooning mode theory is not just a localized theory for a single resonance surface, but a global mode theory.

The ballooning modes have a high toroidal mode number n, unlike interchange modes which can have either a low or high n. Also, the ballooning modes allow normal and geodesic wavelengths to be of the same order $\lambda_\perp \sim \lambda_\wedge$, but both of them are much smaller than the parallel wavelength $\lambda_\parallel$.

In this section, we first derive the high-n mode equations and then describe the ballooning mode representation. The ballooning mode analyses in the so-called s–α model will be given at the end. In addition to the conventional ballooning representation, the representation in configuration space is also described [34].

To derive the high-n mode equations, we note that in the high-n limit both perpendicular components of the momentum equation, (2.6) and (2.7), give rise to the same result

$$\mathbf{B}\cdot\delta\mathbf{B} + \mu_0\delta P = -\left(B^2 + \mu_0\Gamma P\right)\nabla\cdot\boldsymbol{\xi} + B^2\mathbf{B}\cdot\nabla\left(\frac{\mathbf{B}\cdot\boldsymbol{\xi}}{B^2}\right) - 2B^2\boldsymbol{\kappa}\cdot\boldsymbol{\xi} = 0. \quad (2.69)$$

In the lowest order, one has $\nabla\cdot\boldsymbol{\xi}_\perp \sim \xi/R$. This allows us to introduce the so-called stream function $\delta\varphi$, such that $\boldsymbol{\xi}_\perp = \mathbf{B}\times\nabla\delta\varphi/B^2$ [33]. The vorticity equation (2.9) then becomes,

$$\mathbf{B}\cdot\nabla\frac{1}{B^2}\nabla\cdot\left(B^2\nabla_\perp\frac{\mathbf{B}\cdot\nabla\delta\varphi}{B^2}\right) + \nabla\cdot\left(\rho\omega^2\frac{\nabla_\perp\delta\varphi}{B^2}\right)$$
$$+ P'_\psi\nabla\times\frac{\mathbf{B}}{B^2}\cdot\nabla\left(\frac{\mathbf{B}\times\nabla\psi}{B^2}\cdot\nabla\delta\varphi\right) + \Gamma P\nabla\times\frac{\mathbf{B}}{B^2}\cdot\nabla\nabla\cdot\boldsymbol{\xi} = 0. \quad (2.70)$$

Meanwhile, the parallel equation of motion, (2.8), can be reduced to

$$\Gamma P\mathbf{B}\cdot\nabla\left(\frac{1}{B^2}\mathbf{B}\cdot\nabla\nabla\cdot\boldsymbol{\xi}\right) + \rho_m\omega^2\frac{B^2+\mu_0\Gamma P}{B^2}\nabla\cdot\boldsymbol{\xi}$$
$$= \rho_m\omega^2\frac{2\mathbf{B}\times\boldsymbol{\kappa}}{B^2}\cdot\nabla\delta\varphi, \quad (2.71)$$

where (2.69) has been used to express $\nabla\cdot\boldsymbol{\xi}_\perp$.

The key success of the ballooning mode theory lies in the so-called ballooning representation [26, 35]. Here, we outline its physical basis and formulation, in particular to explain the equivalence of the two types of representations in [35] and [26]. Figure 2.2 shows the $n = 10$ modes in the ITER equilibrium as computed by the

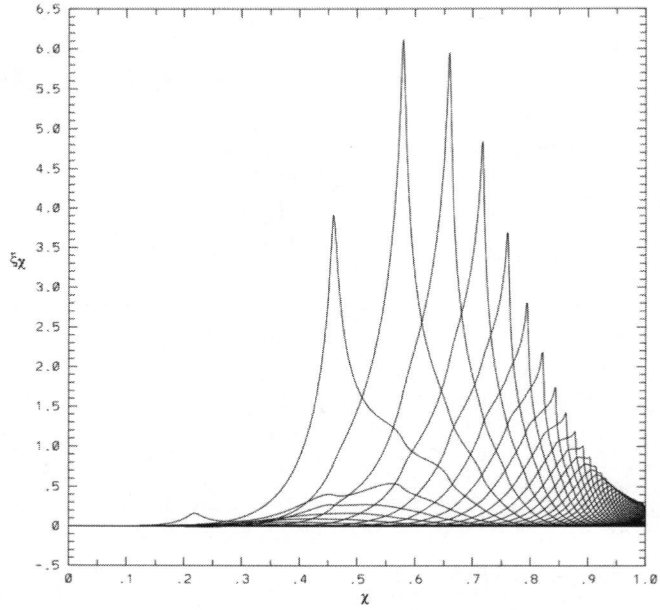

Figure 2.2. The $n = 10$ modes in the ITER configuration, as computed by the AEGIS code. (Reproduced with permission from [12]. Copyright 2006 Elsevier.)

two-dimensional MHD code AEGIS. From this figure one can see that there are similarities between each of the Fourier components. The purpose of the ballooning mode representation is to tap this similarity in order to reduce the two-dimensional problem into a one-dimensional problem. The principle is that in the lowest order each of the poloidal Fourier harmonics is assumed to be of the same shape. The overall envelop is then constructed in the next order. In this section only the lowest order theory is described. One can find the first order theories in [26, 33, 34].

In the tokamak geometry one can introduce the following Fourier decomposition for a perturbed field:

$$\delta\varphi(nq, \theta, \zeta) = \sum_{m=-\infty}^{+\infty} \delta\varphi_m(nq) \exp\{-in\zeta + m\theta\}. \tag{2.72}$$

To describe the ballooning mode representation we have used nq as the flux surface label. This is allowed for the region with a monotonic q profile, in which the ballooning representation applies. For high-n modes the distance between nearby mode rational surfaces is of the order $1/n$ and is much smaller than the equilibrium scale length. Therefore, in the lowest order one can neglect the spatial variance of the equilibrium quantities and require the mode poloidal Fourier harmonics to have the so-called ballooning invariance:

$$\delta\varphi_m(nq) = \delta\varphi(nq - m). \tag{2.73}$$

Using this invariance property, the Fourier decomposition in (2.72) can be expressed as

$$\delta\varphi(nq, \theta, \zeta) = \sum_{m=-\infty}^{+\infty} \delta\varphi(nq - m) \exp\{-in\zeta + m\theta\}. \tag{2.74}$$

One can further introduce the Laplace transform

$$\delta\varphi(nq) = \frac{1}{2\pi} \int_{-\infty}^{+\infty} \delta\varphi(\eta) \exp\{inq\eta\} \, \mathrm{d}\eta. \tag{2.75}$$

Using this transform 2.74 can be written as

$$\delta\varphi(nq, \theta, \phi) = \frac{1}{2\pi} \exp\{-in\zeta\} \int_{-\infty}^{+\infty} \delta\varphi(\eta) \sum_{m} \exp\{im(\theta - \eta)\} \, \mathrm{d}\eta. \tag{2.76}$$

Noting that

$$\frac{1}{2\pi} \sum_{m=-\infty}^{+\infty} \exp\{im(\theta - \eta)\} = \sum_{j=-\infty}^{+\infty} \delta(\eta - \theta - j2\pi),$$

equation (2.76) can be transformed to

$$\delta\varphi(nq, \theta, \zeta) = \sum_{j=-\infty}^{+\infty} \delta\varphi(\theta + j2\pi) \exp\{-in(\zeta - q(\theta + j2\pi))\}. \tag{2.77}$$

This indicates that we can represent the high-n modes at a reference surface as

$$\delta\varphi(nq, \theta, \zeta) = \delta\varphi(\theta) \exp\{-in\beta\}, \tag{2.78}$$

without considering periodicity requirements. Here, $\beta \equiv \zeta - q\theta$. The periodic eigenfunction can always be formed through the summation in (2.77). This representation characterizes the most important feature of ballooning modes in a plasma torus: the perpendicular wave number is much larger than the parallel one, $k_\perp \gg k_\parallel$. This reduction shows the equivalence of the two kinds of representations in (2.74) and (2.77) [26, 35]. The uniqueness and inversion of the ballooning mode representation were proved in [36].

With the ballooning mode representation described above, we can proceed to reduce the ballooning mode equation. It is convenient to use the so-called Celbsch coordinates (ψ, β, θ) to construct the equations. In these coordinates $\nabla_\perp \to -in\nabla\beta$ and $\mathbf{B} \cdot \nabla = \chi'(\partial/\partial\theta)$. Applying (2.78) to (2.70) and (2.71) and employing high-n ordering, one can obtain the following coupled ballooning mode equations

$$\chi' \frac{\partial}{\partial\theta}\left(|\nabla\beta|^2 \chi' \frac{\partial}{\partial\theta}\delta\varphi\right) + P'\nabla \times \frac{\mathbf{B}}{B^2} \cdot \nabla\beta\delta\varphi + \Gamma P \nabla \times \frac{\mathbf{B}}{B^2} \cdot \nabla\beta\delta\Xi$$

$$+ \frac{\omega^2}{\omega_A^2} |\nabla\beta|^2 \delta\varphi = 0, \tag{2.79}$$

$$\Gamma P\chi' \frac{\partial}{\partial \theta}\left(\frac{1}{B^2}\chi' \frac{\partial}{\partial \theta}\delta\Xi\right) + \rho_m \omega^2 \frac{B^2 + \mu_0 \Gamma P}{B^2}\delta\Xi = \rho_m \omega^2 \frac{2\mathbf{B} \times \boldsymbol{\kappa}}{B^2} \cdot \nabla\beta\delta\varphi, \quad (2.80)$$

where $\delta\Xi = \mathrm{i}\nabla \cdot \boldsymbol{\xi}/n$. These two equations are the coupled second order differential equations. The derivatives here are along a reference magnetic field line labeled by ψ and β. The boundary conditions are $\delta\varphi$, $\delta\Xi \to 0$ at $\theta \to \pm\infty$ to guarantee the convergence of the Laplace transform in (2.75).

In studying the ballooning stability at a finite beta equilibrium, the so-called steep pressure gradient equilibrium model is often used [37, 38]. In this model, the finite beta modification is only taken into account for magnetic shear, while other parameters remain at their low beta values. This model has been proved to be successful for ballooning mode studies. Here, we outline the formulation in [27]. Noting that $\beta = q\theta - \zeta$, one can find that the magnetic shear effect resides at the quantity $\nabla\beta$ in the ballooning mode equations (2.79) and (2.80). From (1.19) one can prove that

$$\nabla\beta = \Lambda_s \nabla\chi + \frac{\mathbf{B} \times \nabla\chi}{|\nabla\chi|^2}, \quad (2.81)$$

where Λ_s is the so-called shear parameter and can be obtained by applying operator $\mathbf{B} \times \nabla\chi \cdot \nabla \times \cdots$ on (2.81),

$$\chi' \frac{\mathrm{d}\Lambda_s}{\mathrm{d}\theta} = -\frac{\mathbf{B} \times \nabla\chi \cdot \nabla \times (\mathbf{B} \times \nabla\chi)}{|\nabla\chi|^4}. \quad (2.82)$$

One needs to determine the finite beta modification to the shear parameter Λ_s. It is assumed that $\chi = \chi_0 + \chi_1$ and $\beta = \beta_0 + \beta_1$, where χ_0 and β_0 are the low beta values and χ_1 and β_1 represent the finite beta modifications. The linearized Ampere's law can be written as follows:

$$\nabla \times \left(\nabla\chi_0 \times \nabla\beta_1 + \nabla\chi_1 \times \nabla\beta_0\right) = \mathbf{J} = \frac{\partial P}{\partial \chi}\left(2\lambda\nabla\chi_0 \times \nabla\beta_0 + \frac{\mathbf{B}_0 \times \nabla\chi_0}{B^2}\right). \quad (2.83)$$

Noting that in applying the curl operation on the left-hand side of this equation, only the gradient component in the $\nabla\chi$-direction needs to be kept, i.e. $\nabla \times \to \nabla\chi_0 \partial/\partial\chi \times$, equation (2.83) can be solved,

$$2P\lambda\nabla\beta_0 + \mathbf{B}_0 P + \nabla Q = \nabla\chi_0 \times \nabla\beta_1 + \nabla\chi_1 \times \nabla\beta_0, \quad (2.84)$$

where ∇Q is the integration factor. Noting that only $\partial\{P, Q\}/\partial\chi$ are large, the divergence of (2.84) gives

$$2\lambda \frac{\partial P}{\partial \chi}\left(\nabla\chi_0 \cdot \nabla\beta_0\right) + \frac{\partial^2 Q}{\partial \chi^2}\left|\nabla\chi_0\right|^2 = 0.$$

Solving this equation, one obtains

$$\frac{\partial Q}{\partial \chi} = -2\lambda P \frac{\nabla\chi \cdot \nabla\beta}{|\nabla\chi|^2}. \quad (2.85)$$

Taking the dot product of (2.84) with $\nabla\beta_0$ gives

$$2P\lambda\left|\nabla\beta_0\right|^2 + \left(\nabla\chi_0\cdot\nabla\beta_0\right)\frac{\partial Q}{\partial\chi} = -\chi'\frac{\partial\beta_1}{\partial\theta},$$

We can remove subscript 0 afterward for brevity. Now substituting (2.85) for $\partial Q/\partial\chi$ and noting that $\partial\beta_1/\partial\chi \equiv \Lambda_{s1}$, one finds that

$$\chi'\frac{\partial\Lambda_{s1}}{\partial\theta} = -2\lambda\frac{\partial P}{\partial\chi}\frac{B^2}{|\nabla\chi|^2}.$$

Therefore, the shear parameter can be evaluated as follows

$$\chi'\frac{\mathrm{d}\Lambda_s}{\mathrm{d}\theta} = -\frac{\mathbf{B}\times\nabla\chi\cdot\nabla\times(\mathbf{B}\times\nabla\chi)}{|\nabla\chi|^4} - 2\lambda\frac{\partial P}{\partial\chi}\frac{B^2}{|\nabla\chi|^2}. \tag{2.86}$$

The second term here gives rise to the finite beta modification to the shear parameter Λ_s in the steep pressure gradient model. The rest of the parameters here and in the ballooning equations (2.79) and (2.80) can be evaluated with their low beta values.

Next, we consider the tokamak equilibrium with a circular cross section and a large aspect ratio (i.e. $1/\varepsilon = R/a \gg 1$). The magnetic field in this model can be expressed as $\mathbf{B} = B_\phi(r)/(1+\varepsilon\cos\theta)\mathbf{e}_\phi + B_\theta(r)\mathbf{e}_\theta$. The shear parameter in (2.86) can be expressed as $\Lambda_s = s(\theta-\theta_\mathrm{k}) - \alpha\sin\theta$. Here, $\alpha = -(2\mu_0 Rq^2/B^2)(\mathrm{d}P/\mathrm{d}r)$, $s = d\ln q/d\ln r$ and θ_k is the integration constant. Therefore, the ballooning equations (2.79) and (2.80) can be reduced to

$$\frac{\mathrm{d}}{\mathrm{d}\theta}\left(\left(1+\Lambda_s^2\right)\frac{\mathrm{d}\delta\varphi}{\mathrm{d}\theta}\right) + \alpha\left(\cos\theta+\Lambda_s\sin\theta\right)\delta\varphi + \frac{2\Gamma RrqP}{B}\left(\cos\theta+\Lambda_s\sin\theta\right)\delta\Xi$$
$$+\frac{\omega^2}{\omega_\mathrm{A}^2}\left(1+\Lambda_s^2\right)\delta\varphi = 0, \tag{2.87}$$

$$\frac{\Gamma P}{R^2q^2}\frac{\partial^2\delta\Xi}{\partial\theta^2} + \rho_m\omega^2\delta\Xi = -\frac{2\rho_m\omega^2}{R^2B_\theta}\left(\cos\theta+\Lambda_s\sin\theta\right)\delta\varphi. \tag{2.88}$$

To further analyze this set of equations it is interesting to consider two limits: the low frequency ($\omega \ll \omega_\mathrm{si}$) and intermediate frequency ($\omega_\mathrm{si} \ll \omega \ll \omega_\mathrm{se}$) limits, where ω_si and ω_se are the ion and electron acoustic frequencies, respectively. The comparable frequency regime ($\omega \sim \omega_\mathrm{si}$) involves the wave–particle resonances and therefore needs a kinetic treatment [39]. In the low frequency limit the second term on the left-hand side of (2.88) can be neglected. It can be proved *a posteriori* that the sound wave coupling in the low frequency regime only enhances the apparent mass. Note further that the inertia term is only important in the outer region $\theta\to\infty$. equation (2.88) can be solved in the outer region limit, yielding

$$\delta\Xi = \frac{2\rho_m q^2\omega^2}{\Gamma PB_\theta}s\theta\sin\theta\delta\varphi. \tag{2.89}$$

Here, it has been noted that the slow variable $s\theta$ can be regarded as constant in the sound wave scale. Inserting (2.89) into (2.87) results in the following single ballooning mode equation

$$\frac{\mathrm{d}}{\mathrm{d}\theta}\left(\left(1 + \Lambda_s^2\right)\frac{\mathrm{d}\delta\varphi}{\mathrm{d}\theta}\right) + \alpha\left(\cos\theta + \Lambda_s \sin\theta\right)\delta\varphi + \frac{\omega^2}{\omega_{\mathrm{A}}^2}\left(1 + 2q^2\right)s^2\theta^2\delta\varphi = 0. \quad (2.90)$$

Here, one can find that sound wave coupling results in the so-called apparent mass effect: i.e. the inertia term (the final term in (2.90)) is enhanced by a factor $(1 + 2q^2)$ [22]. In the kinetic description the $2q^2$ term is further boosted by the so-called small particle speed effect to become of order $2q^2/\sqrt{r/R}$ for the large aspect ratio case [2, 28].

In the marginal stability $\omega^2 = 0$ the ballooning stability can be determined by Newcomb's theorem 5 [29]: the system is unstable, if and only if the solution of (2.61) vanishes at two or more points (or is a small solution at infinity). In [37] and [40] the stability boundaries have been obtained for ballooning modes, see the solid curves in figure 2.3. They found that there are two stability regimes: the first and the second stability regimes. Much theoretical effort has been focused on finding a way to achieve the second stability regime. The dashed curves in figure 2.3 represent the peeling mode stability boundary, which will be discussed in section 2.5.

In the intermediate frequency regime the first term in (2.88) can be neglected and therefore one obtains

$$\delta\Xi = -\frac{2}{R^2 B_\theta}\left(\cos\theta + \Lambda_s \sin\theta\right)\delta\varphi. \quad (2.91)$$

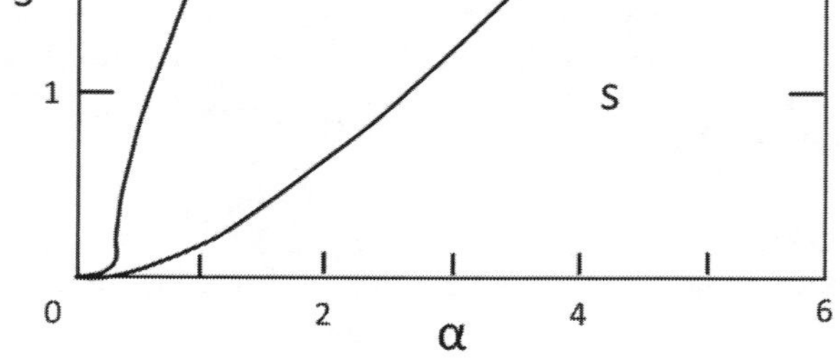

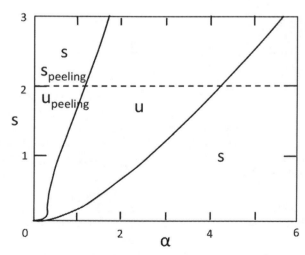

Figure 2.3. The marginal stability boundaries in the shear s and normalized core beta α space. The solid curves represent the ballooning modes; the dashed line represents the peeling modes.

Inserting (2.91) into (2.87) yields [41]

$$\frac{\mathrm{d}}{\mathrm{d}\theta}\left(\left(1+\Lambda_s^2\right)\frac{\mathrm{d}\delta\varphi}{\mathrm{d}\theta}\right)+\alpha\left(\cos\theta+\Lambda_s\sin\theta\right)\delta\varphi-\frac{4\mu_0\Gamma q^2 P}{B^2}\left(\cos\theta+\Lambda_s\sin\theta\right)^2\delta\varphi$$

$$+\frac{\omega^2}{\omega_{\mathrm{A}}^2}\left(1+\Lambda_s^2\right)\delta\varphi=0. \tag{2.92}$$

As shown in section 2.6, the sound wave coupling term (the third term) can give rise to the so-called second harmonic TAE in the circular cross section case [3].

Next, let us introduce an alternative ballooning mode representation [34]. The conventional ballooning representation in (2.77) leads to differential equations for a single Fourier component in the infinite θ domain. There exists the other possibility of considering all the Fourier components but only in a single resonance interval, $x_0 \rightarrow x_0 + \Delta$, as shown in figure 2.4. Then the ballooning invariance in (2.73) is applied to construct the boundary conditions to link the different Fourier components. Suppose that the maximum and minimum Fourier components to be considered are $m_{\max}$ and $m_{\min}$. According to the invariance property in (2.73), the Fourier components inside this range are required to satisfy the following connection conditions

$$\delta\varphi_m\left(x_0+\Delta\right)=\mathrm{e}^{-\mathrm{i}\theta\mathrm{k}}\,\delta\varphi_{m+1}\left(x_0\right), \tag{2.93}$$

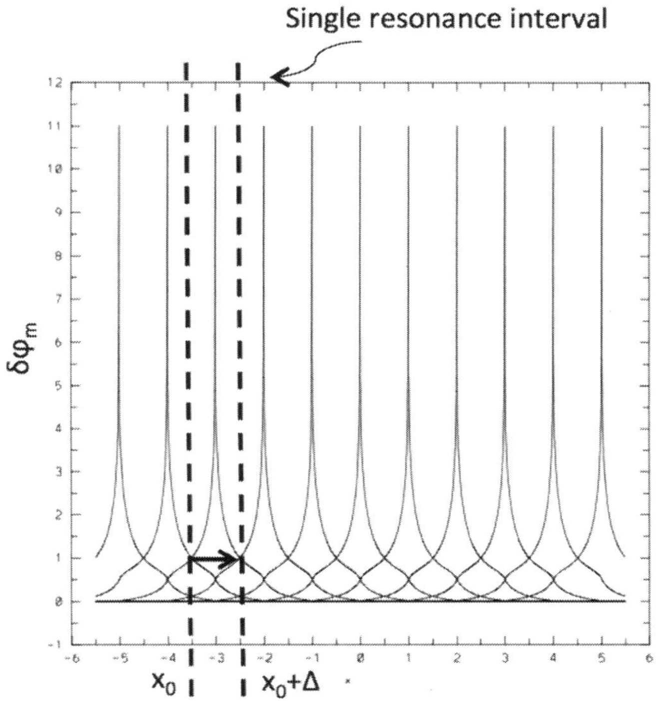

Figure 2.4. The ballooning mode representation in configuration space. (Reproduced with permission from [34]. Copyright 2012, AIP Publishing LLC.)

$$\frac{d}{dx}\delta\varphi_m(x_0 + \Delta) = e^{-i\theta k}\frac{d}{dx}\delta\varphi_{m+1}(x_0). \tag{2.94}$$

As for the farthest sidebands ($m_{\min}$ and $m_{\max}$) one can require

$$\delta\varphi_{m\max}(x_0 + \Delta) = 0, \tag{2.95}$$

$$\delta\varphi_{m\min}(x_0) = 0, \tag{2.96}$$

or small solutions at the singular surfaces. The boundary conditions (2.93)–(2.96) fully determine the eigenvalue problem in the interval $x_0 \to x_0 + \Delta$. Figure 2.4 is an example of the eigenfunction of the lowest order in the s–α model, computed using the one-dimensional ballooning mode formalism in configuration space [34].

The computational resources used to compute multiple Fourier components in a single interval in this configurational space ballooning formulation are equivalent to those used to compute a single Fourier component in the expanded infinite θ space in the conventional ballooning formalism. The single radius interval code can have a more apparent poloidal transport picture than a flux tube code. In addition, as will be discussed in the following section, the configuration space ballooning formalism can be generalized to treat the free boundary ballooning mode problem, and even to include pedestal physics.

2.5 The peeling and free boundary ballooning modes

In this section, we discuss the peeling and free boundary ballooning modes [34, 42, 43]. The peeling and peeling–ballooning modes have attracted much attention recently in the field due to their link to the plasma edge instabilities [44]. We first describe the peeling mode theory and then the free boundary ballooning mode theory.

The most notable feature of the peeling and free boundary ballooning modes lies in the fact that they are external modes, i.e. their resonance surfaces lie in the vacuum region. We note that the infinity is also regarded as the singular surface for individual Fourier harmonics. This feature can make a big difference. Figure 2.5 shows the free boundary ballooning modes with the last rational surfaces in the vacuum and plasma regions, respectively. They are computed with the same s–α equilibrium model as the internal ballooning modes in figure 2.4. However, comparing figures 2.4 and 2.5 one can see that their eigenmode behaviors differ dramatically. The envelopes of the eigenmodes in figure 2.5 are not due to radial equilibrium parameter changes—s and α are constant in the equilibrium model. The envelopes in figure 2.5 are due to lowest order effects. The conventional ballooning invariance breaks down at the plasma edge. The Fourier components with their resonance surfaces in the plasma region and those with their resonance surfaces in the vacuum region behave completely differently. The early efforts to obtain a free boundary ballooning mode representation using the next order theory did not succeed, since the conventional ballooning invariance breaks down in the lowest order.

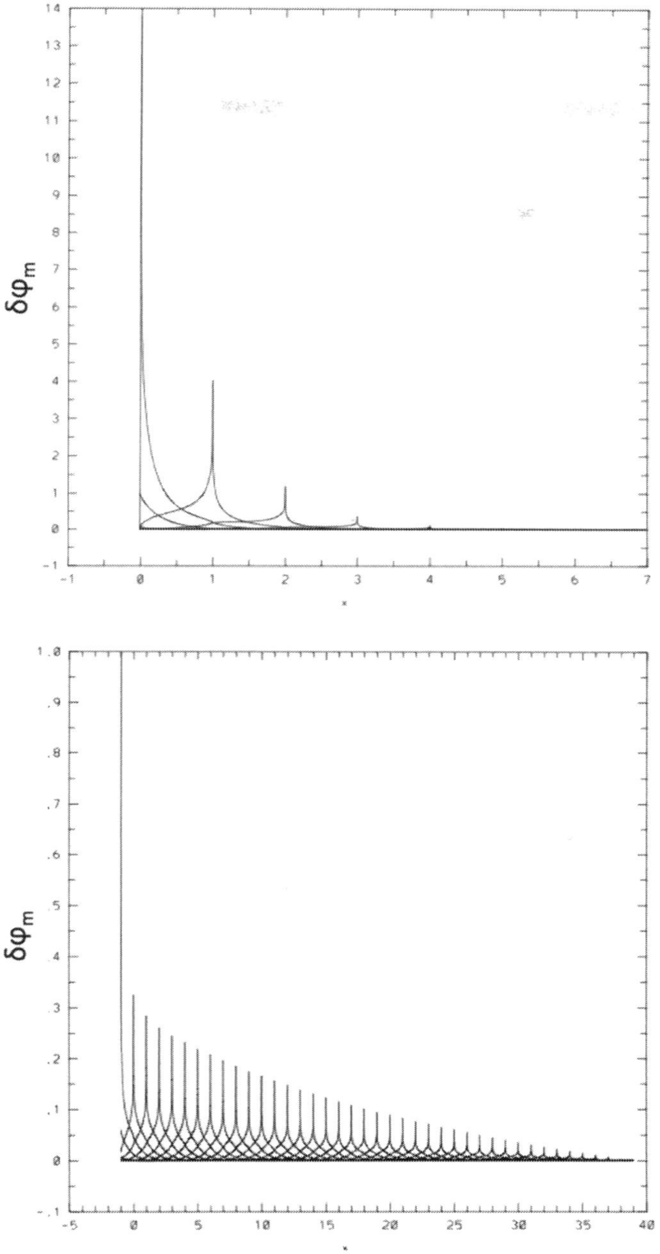

Figure 2.5. The free boundary ballooning modes with the last rational surfaces in the vacuum and plasma regions, respectively. (Reproduced with permission from [34]. Copyright 2012, AIP Publishing LLC.)

Let us use the mode behavior in the singular layer in (2.66) to discuss the underlying mathematical physics. In the discussion we regard the infinity from a reference surface to be the singular surface as well. Therefore, each of the harmonics has two singular surfaces on each side of the rational surfaces: the rational surface and infinity. Inside the plasma only the small solution is allowed at the singular surface. If the Mercier index $-D_{\mathrm{I}} < 0$, the mode becomes unstable before reaching the other singular surface. If $-D_{\mathrm{I}} > 0$ between the two singular surfaces, the internal mode stability depends on whether the solution on the other singular surface also inside the plasma is large or not, i.e. the sign $\pm$ in (2.66). In the case of the rational surface, the solution is referred to as the large solution if the negative sign is chosen and vice versa for the infinite singularity case. For external modes, however, one of the two singular surfaces lies in the vacuum region. The large solution at that singular surface is acceptable in this case. Note as well that the vacuum energy is negligible for localized or high-n modes. One can conclude that the external modes, the peeling or free boundary ballooning modes, are more unstable. It is due to this that the last Fourier component peaks in figure 2.5. Therefore, the Fourier components with their mode rational surfaces lying in the vacuum region are completely different from those with their mode rational surfaces lying in the plasma region. They cannot be translationally invariant as given in (2.73).

Now, let us describe the peeling mode theory. Since peeling modes are the external modes, one needs to consider the matching condition between the plasma and vacuum solutions. As discussed in many MHD books, these matching conditions for the case without a surface plasma current are [4]: (1) the tangential magnetic perturbation ($\delta \mathbf{B}_{\mathrm{t}}$) should be continuous; and (2) the total magnetic and thermal forces ($\mathbf{B} \cdot \delta \mathbf{B} + \delta P$) should balance across the plasma–vacuum interface. It can be proved that for localized modes the vacuum contribution is of the order ϵ^2 and therefore can be neglected [42]. Consequently, the boundary condition becomes such that the total magnetic and thermal forces on the plasma side of plasma–vacuum interface should vanish, i.e.

$$[\mathbf{B} \cdot \delta \mathbf{B} + \delta P]_{x=b} = 0. \tag{2.97}$$

where b is the coordinate of the plasma–vacuum interface, relative to the rational surface.

This condition can be reduced by projecting the momentum equation in the direction $\nabla V \times \nabla \theta$ and then performing the surface averaging. Since the procedure is similar to the derivation of the Mercier condition in section 2.3 and the resistive MHD theory in section 3.1, we omit the tedious details and just provide the key steps (also partially referring to [45]). Noting that

$$\nabla V \times \nabla \theta \cdot \delta \mathbf{j} \times \mathbf{B} = -\chi' \frac{\partial}{\partial \theta}(\cdots) + \chi' \frac{\partial}{\partial u}\left(\frac{\Lambda}{\Xi} \frac{|\nabla V|^2}{B^2} x v^{(1)} - \frac{\tau^{(1)} + \tau^{(2)}}{\Xi} \right),$$

$$\nabla V \times \nabla \theta \cdot \mathbf{j} \times \delta \mathbf{B} = J' b^{(2)},$$

$$\nabla V \times \nabla\theta \cdot \nabla\delta P = \frac{\chi'}{\Xi}\frac{\partial}{\partial u}\left(\delta P^{(1)} + \delta P^{(2)}\right),$$

$$-i\rho_m\omega^2\nabla V \times \nabla\theta \cdot \boldsymbol{\xi} = \frac{\rho_m\omega^2\Xi}{i\alpha p'^2}\left(B^2\sigma^2 - \frac{\left\langle B^2\sigma\right\rangle}{\left\langle B^2\right\rangle^2}B^2 + \frac{p'^2|\nabla V|^2}{B^2}\right)\frac{\partial\xi^{(1)}}{\partial x},$$

and also $\tau^{(1)} + \delta P^{(1)} = 0$, (2.49) and (2.56), one has

$$\begin{aligned}\langle \mathbf{B}\cdot\delta\mathbf{B} + \delta P\rangle &= \left\langle \tau^{(2)} + \delta P^{(2)}\right\rangle \\ &= -\frac{\Lambda^2}{\left\langle B^2/|\nabla V|^2\right\rangle}\left[\left(x^2 - M\omega^2\right)\frac{\mathrm{d}\xi}{\mathrm{d}x} + \left(\Delta + \frac{1}{2}\right)x\xi\right]. \end{aligned} \tag{2.98}$$

where

$$\Delta = \frac{1}{2} - \Lambda^{-1}\left\langle\frac{B^2\sigma}{|\nabla V|^2}\right\rangle.$$

In passing, we note that, using the projections of the momentum equations on three directions directly, $\nabla V \times \nabla u$, $\nabla u \times \nabla\theta$ and $\nabla V \times \nabla\theta$, one can also derive the Mercier criterion [23].

Substituting (2.99) into (2.97), one obtains the following necessary and sufficient stability condition for the peeling modes

$$\left[\frac{x^2}{2}\left(\xi^*\frac{\mathrm{d}\xi}{\mathrm{d}x} + \xi\frac{\mathrm{d}\xi^*}{\mathrm{d}x}\right) + \left(\Delta + \frac{1}{2}\right)x\,|\xi|^2\right]_{x=b} > 0. \tag{2.99}$$

Note that the stability condition (2.99) can be alternatively obtained by the minimization of energy principle.

One can derive the peeling mode stability criterion by inserting (2.66) into (2.99) [43]. In the derivation of the peeling stability criterion we assume the system to be Mercier stable, i.e. $-D_{\mathrm{I}} > 0$. For the case with $\Delta > 0$ we assume that the rational surface resides inside the plasma region, so that $b > 0$. In this case the stability condition becomes

$$\sqrt{-D_{\mathrm{I}}} + \Delta > 0. \tag{2.100}$$

For the case with $\Delta < 0$ we assume that the rational surface resides outside the plasma region, so that $b < 0$. In this case the stability condition becomes

$$\sqrt{-D_{\mathrm{I}}} - \Delta > 0. \tag{2.101}$$

Note that $-D_{\mathrm{I}} \equiv \Delta^2 - \Lambda_{\mathrm{p}}$, where $\Lambda_{\mathrm{p}} = \Lambda^{-2}\langle \mathbf{J}^2\,|\nabla V|^2 + I'\psi' - J'\chi''\rangle\langle B^2\,|\nabla V|^{-2}\rangle$. Therefore, both cases, (2.100) and (2.101), give rise to the same stability criterion for the peeling mode: $\Lambda_p < 0$. This is more stringent than the Mercier criterion. In the s–α equilibrium model the peeling stability condition becomes $s > 2$ in figure 2.4 [46].

Next, let us discuss the free boundary ballooning modes. The free boundary ballooning modes can be computed using two-dimensional codes, such as the ELITE and AEGIS codes [12, 44]. Here, we focus on the one-dimensional formalism, i.e. the free boundary ballooning mode representation [34]. Unlike the peeling mode, which only considers the single resonant Fourier component with coupling to two nearby harmonics, the free boundary ballooning modes take into account the multiple-mode coupling physics. They are nonlocal, but can be mathematically treated as local using a suitable representation.

As discussed at the beginning of this section, the conventional ballooning invariance in (2.73) becomes invalid at the plasma edge. However, the ordering separation for high-n modes, i.e. the distance between the nearby mode rational surfaces is much smaller than the equilibrium scale length, remains unchanged. Therefore, the possibility of tapping this ordering separation remains open and one simply needs to figure out the mathematical formulation.

To cope with this mathematical problem, the edge coordinate system in figure 2.6 is introduced. The intervals are labeled from '0' to '$k_{\max}$' from the interval adjacent to the plasma–vacuum interface to the farthest interval in the plasma core. For specificity we assume that the safety factor increases monotonically from the plasma core to the edge. Note that the solution of a set of linear differential equations contains two steps: finding the independent solutions and determining the boundary conditions by the linear combination of independent solutions. The conventional ballooning invariance requires both the independent solutions and boundary conditions to be invariant. The requirement for the boundary conditions makes the conventional ballooning mode representation inapplicable to the edge region. In fact, the requirement for boundary value invariance is not part of the high-n mode scale separation and is therefore unnecessary. The invariance of the independent solutions is more fundamental. Mathematically, this invariance can be expressed as follows

$$^j\delta\varphi_{m\mp 2k}(x \pm k) = {}^j\delta\varphi_m(x), \tag{2.102}$$

where the index j labels the independent solutions and k indicates the regions as shown in figure 2.6. The difference from the conventional ballooning invariance in

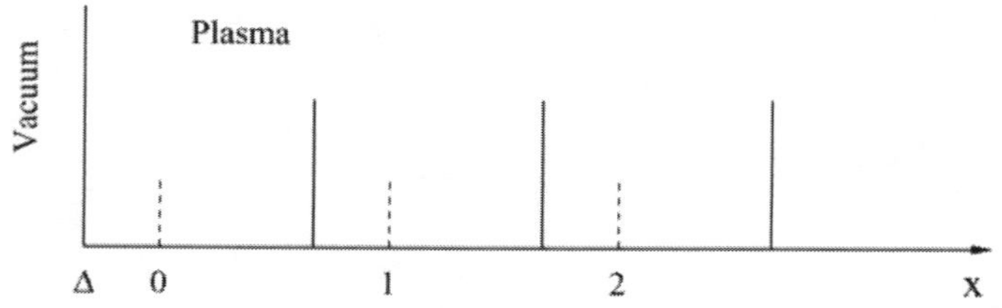

Figure 2.6. The coordinate system, in which the vertical dashed lines represent the resonance surfaces and the vertical solid lines denote the region boundaries. (Reproduced with permission from [34]. Copyright 2012, AIP Publishing LLC.)

(2.73) lies in that the invariance here is related to the independent solutions, instead of the Fourier components of the perturbation.

When the independent solutions are obtained in a single interval, say the $k=0$ region in figure 2.6, one in fact obtains the independent solutions in all the intervals by applying the invariance in (2.102). With the independent solutions known for all intervals, the eigenvalue problem can be constructed through fitting the linear combination of the independent solution in each interval to the continuity conditions between the nearby intervals and boundary conditions at the plasma core and vacuum–plasma interface. The eigenmodes in figure 2.5 are computed using this one-dimensional formalism [34]. The peeling mode stability condition, $s > 2$ for the s–α equilibrium model, is recovered in the free boundary ballooning mode calculation.

The fundamental principle for constructing the free boundary ballooning mode representation can be extended to deal with the multiple region problem, for example pedestal physics. In this case, one can construct two sets of independent solutions for an interval in the core and pedestal regions, respectively, and then extend the independent solution to all intervals for the core and pedestal regions. Finally, the linear combinations of the two set of independent solutions are required to satisfy the continuity and boundary conditions. In this way the two-dimensional pedestal problem can be transformed into a quasi-one-dimensional eigenvalue problem.

2.6 The toroidal Alfvén eigenmodes

In this section, we describe the toroidal Alfvén eigenmode (TAE) theory. The global theory has been discussed in sections 2.1 and 2.2. In section 2.1 we found that the toroidal geometry can induce the first frequency gap so that the standing wave becomes an eigenmode, i.e. the so-called TAE [47, 48]. In section 2.2 we discussed the global TAE computation. The typical $n=1$ TAE eigenfunction in equilibrium with a circular cross section and a large aspect ratio is shown in figure 2.7 from global calculations. In this section we describe the analytical theory for TAEs.

In sections 2.3 and 2.4 we found that the interchange and ballooning modes are characterized by having only a single dominant resonant mode at each resonance surface. In particular their resonance surfaces coincide with the mode rational surface where $m - nq = 0$. The TAEs are different. They involve two-mode coupling. In particular, the first TAEs are centered at the surface where $q = \left(m_0 + 1/2\right)/n$. Two neighboring Fourier modes (m_0 and $m_0 + 1$) propagate roughly with the same speed $v_A/2Rq$ but in the opposite directions. The two harmonics form a standing wave. Indeed, in figure 2.7 one can see that the $m=1$ and 2 harmonics resonate at the $q=1.5$ surface. In the second TAE case, although they have the same mode resonance surface as the interchange or ballooning modes, the $m_0 \pm 1$ mode couplings are involved in forming the standing second TAEs. The frequency gap for the second TAEs in the circular cross section case is due to the plasma compressibility effect [3]. This is different from the elliptically induced Alfvén eigenmodes in [49].

We use the model tokamak equilibrium with a circular cross section, low beta and a large aspect ratio (i.e. $1/\epsilon = R/a \gg 1$) to explain the physics picture of two-mode coupling. There is a review paper on TAEs [50]. Here, we describe the local dispersion

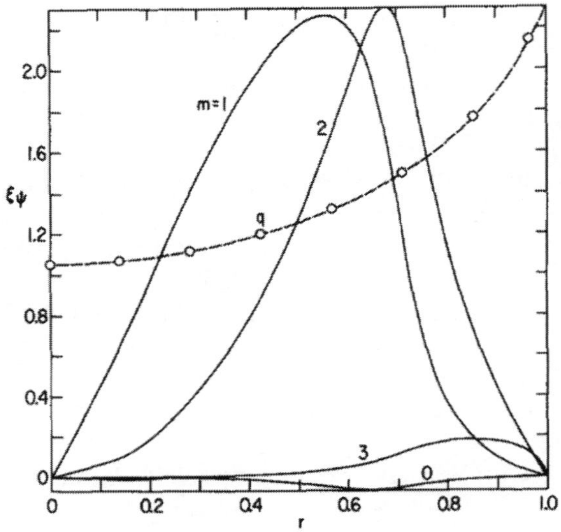

Figure 2.7. The $n = 1$ TAE with a frequency in the gap shown in figure 2.1. (Reproduced with permission from [15]. Copyright 1986, AIP Publishing LLC.)

relation for even and odd modes and explain the second TAEs together with the first TAEs as in [1]. The magnetic field in this model equilibrium can be expressed as $\mathbf{B} = B_\phi(r)/(1 + \epsilon \cos\theta)\mathbf{e}_\phi + B_\theta(r)\mathbf{e}_\theta$. Since the frequency of compressional Alfvén modes is much larger than the shear Alfvén mode frequency, they are decoupled from the TAEs. Therefore, we can use (2.70) and (2.71) as the starting equations for our TAE investigation. Noting that the Alfvén frequency is much larger than the sound wave frequency, the first term in (2.71) can be dropped. Adopting the Fourier decomposition in (2.72), the sound wave equation (2.71) becomes

$$\mathrm{i}(\nabla \cdot \boldsymbol{\xi})_m = \frac{1}{BR}\left(\frac{\mathrm{d}\delta\varphi_{m+1}}{\mathrm{d}r} - \frac{\mathrm{d}\delta\varphi_{m-1}}{\mathrm{d}r}\right).$$

Using this solution for $\nabla \cdot \boldsymbol{\xi}$, (2.70) can be reduced to [6]

$$\begin{aligned}
&\frac{\mathrm{d}}{\mathrm{d}r}\left[r^3\left(\frac{1}{q} - \frac{n}{m}\right)^2 \frac{\mathrm{d}}{\mathrm{d}r}E_m\right] - \frac{\mathrm{d}}{\mathrm{d}r}\left(r^3\frac{R^2\omega^2}{m^2 v_\mathrm{A}^2}\frac{\mathrm{d}}{\mathrm{d}r}E_m\right) - \epsilon\left(r^3\frac{R^2\omega^2}{m^2 v_\mathrm{A}^2}\frac{\mathrm{d}^2}{\mathrm{d}r^2}E_{m+1}\right)\\
&- \epsilon\left(r^3\frac{R^2\omega^2}{m^2 v_\mathrm{A}^2}\frac{\mathrm{d}^2}{\mathrm{d}r^2}E_{m-1}\right) + \frac{\Gamma P r^3}{B^2 m^2}\frac{\mathrm{d}^2 E_{m+2}}{\mathrm{d}r^2} + \frac{\Gamma P r^3}{B^2 m^2}\frac{\mathrm{d}^2 E_{m-2}}{\mathrm{d}r^2} - wE_m\\
&- \frac{\alpha r^2}{2mq^2}\frac{\mathrm{d}E_{m+1}}{\mathrm{d}r} + \frac{\alpha r^2}{2mq^2}\frac{\mathrm{d}E_{m-1}}{\mathrm{d}r} - \frac{\alpha r}{2q^2}E_{m+1} + \frac{\alpha r}{2q^2}E_{m-1} = 0,
\end{aligned} \tag{2.103}$$

where $E_m = \delta\varphi_m/r$, $v_\mathrm{A}^2 = B_0^2/\rho_m$, B_0 denotes the magnetic field at the magnetic axis and w is introduced to denote the rest magnetic well effects.

We need to examine the singular layer physics for TAEs. In the singular layer only the terms that contain the second order derivative with respect to r need to be taken into account. From the first six terms in (2.103) one can see that the second TAEs (coupling of E_{m-1} and E_{m+1}) have a structure similar to the first TAEs (coupling of E_m and E_{m+1}). The first TAE coupling is due to the finite value of the aspect ratio, while the second TAE coupling is due to the finite beta value. For brevity we focus on the first TAE case. Denoting $\omega_0 = \omega_A/2, q_0 = (m + 1/2)/n, \delta\omega = \omega - \omega_0$ and $\delta q = q - q_0$, the singular layer equations describing the coupling of m and $m + 1$ harmonics become

$$\frac{\partial}{\partial \delta q}\left[\frac{\delta\omega}{2\omega_0} - \left(1 - \frac{1}{2m+1}\right)n\delta q\right]\frac{\partial}{\partial \delta q}\delta\varphi_m = -\frac{\varepsilon}{4}\frac{\partial^2}{\partial \delta q^2}\delta\varphi_{m+1},$$

$$\frac{\partial}{\partial \delta q}\left[\frac{\delta\omega}{2\omega_0} + \left(1 + \frac{1}{2m+1}\right)n\delta q\right]\frac{\partial}{\partial \delta q}\delta\varphi_{m+1} = -\frac{\varepsilon}{4}\frac{\partial^2}{\partial \delta q^2}\delta\varphi_m.$$

Introducing even and odd modes, $\delta\varphi_\pm = \delta\varphi_m \pm \delta\varphi_{m+1}$, these two equations can be reduced to

$$\frac{\partial}{\partial \delta q}\left(\frac{\delta\omega}{2\omega_0} + \frac{1}{2m_0+1}n\delta q\right)\frac{\partial}{\partial \delta q}\delta\varphi_+ - \frac{\partial}{\partial \delta q}n\delta q\frac{\partial}{\partial \delta q}\delta\varphi_- = -\frac{\varepsilon}{4}\frac{\partial^2}{\partial \delta q^2}\delta\varphi_+,$$

$$\frac{\partial}{\partial \delta q}\left(\frac{\delta\omega}{2\omega_0} + \frac{1}{2m_0+1}n\delta q\right)\frac{\partial}{\partial \delta q}\delta\varphi_- - \frac{\partial}{\partial \delta q}n\delta q\frac{\partial}{\partial \delta q}\delta\varphi_+ = \frac{\varepsilon}{4}\frac{\partial^2}{\partial \delta q^2}\delta\varphi_-.$$

Integrating over δq once, one obtains

$$\mathcal{D}\begin{pmatrix}\dfrac{\partial\delta\varphi_+}{\partial\delta q}\\[2ex] \dfrac{\partial\delta\varphi_-}{\partial\delta q}\end{pmatrix} \equiv \begin{pmatrix}\dfrac{\delta\omega}{2\omega_0} + \dfrac{1}{2m_0+1}n\delta q + \dfrac{\epsilon}{4} & -n\delta q\\[2ex] -n\delta q & \dfrac{\delta\omega}{2\omega_0} + \dfrac{1}{2m_0+1}n\delta q - \dfrac{\epsilon}{4}\end{pmatrix}\begin{pmatrix}\dfrac{\partial\delta\varphi_+}{\partial\delta q}\\[2ex] \dfrac{\partial\delta\varphi_-}{\partial\delta q}\end{pmatrix}$$
$$= \begin{pmatrix}A_+\\ A_-\end{pmatrix}, \qquad (2.104)$$

where $\mathcal{D}$ is a 2×2 matrix and $A_\pm$ are the integration constants. Integration of (2.104) across the TAE singular layer (i.e. from δq^- to δq^+) gives the TAE dispersion relation

$$\left(\frac{\delta\varphi_+\big|_{\delta q^-}^{\delta q^+}}{A_+} - \int_{\delta q^-}^{\delta q^+}\frac{\mathcal{D}_{22}\mathrm{d}\delta q}{\det|\mathcal{D}|}\right)\left(\frac{\delta\phi_-\big|_{\delta q^-}^{\delta q^+}}{A_-} - \int_{\delta q^-}^{\delta q^+}\frac{\mathcal{D}_{11}\mathrm{d}\delta q}{\det|\mathcal{D}|}\right)$$
$$= \left(\int_{\delta q^-}^{\delta q^+}\frac{\mathcal{D}_{12}\mathrm{d}\delta q}{\det|\mathcal{D}|}\right)^2. \qquad (2.105)$$

Here, $\mathcal{D}_{ij}$ represent the elements of matrix $\mathcal{D}$ and the two parameters $\Delta_\pm \equiv \delta\varphi_\pm|_{\delta q^-}^{\delta q^+}/A_\pm$ are determined by the outer solutions to the left and right of the singular layer. As soon as $\Delta_\pm$ are computed from the outer regions, equation (2.105) can be used to determine the eigenfrequency.

The denominators of integrations in (2.105) involve the determinant det $|\mathcal{D}|$. The singularity emerges at det $|\mathcal{D}| = 0$. In this case the Landau integration orbit needs to be used, as in the case for particle–wave resonances, and continuum damping occurs [51]. The so-called first TAE frequency gap, in which eigenmodes can exit without continuum damping, can therefore be determined by the condition det $|\mathcal{D}| = 0$, i.e.

$$\left(\frac{\delta\omega}{2\omega_0} + \frac{1}{2m+1}n\delta q + \frac{\varepsilon}{4}\right)\left(\frac{\delta\omega}{2\omega_0} + \frac{1}{2m+1}n\delta q - \frac{\varepsilon}{4}\right) = n^2\delta q^2.$$

Its solution is

$$\left[1 - \left(\frac{1}{2m_0+1}\right)^2\right]n^2\delta q = \frac{\delta\omega}{2\omega_0(2m_0+1)} \pm \sqrt{\left(\frac{\delta\omega}{2\omega_0}\right)^2 - \frac{\varepsilon^2}{16}\left[1 - \frac{1}{(2m+1)^2}\right]}.$$

To exclude a real δq solution for det $|\mathcal{D}| = 0$, the mode frequency must fall in the gap between $\delta\omega_\pm$, i.e. $\omega_- < \omega < \omega_+$, where

$$\delta\omega_\pm = \pm\frac{\varepsilon}{2}\omega_0\sqrt{1 - 1/(2m+1)^2}.$$

One can obtain the gap width $\Delta\omega = \delta\omega_+ - \delta\omega_- = \varepsilon\omega_0\sqrt{1 - 1/(2m+1)^2}$. The first TAEs are the Alfvén eigenmodes with a frequency inside this gap. They are marginally stable and tend to be excited by the resonances with energetic particles. Note that the gap width is proportional to ϵ. The TAE gap vanishes in the cylinder limit, as shown by the dashed curves in figure 2.1. Therefore, the existence of TAEs is due to toroidal effects. Also, we note that the dispersion relation, (2.105), allows two types of TAEs, even and odd types ($\varphi_\pm$), depending on the values of $\Delta_\pm$. Figure 2.1 is the even mode example. In the odd mode case the $m = 1$ and $m = 2$ harmonics have different signs at the singular surface.

We have discussed the first TEA theory through the coupling of two neighboring harmonics. In a similar way one can also develop the second TAE theory by considering the coupling of $m \pm 1$ harmonics [3]. If the FLR effects are taken into consideration, the Alfvén-type singularities can be resolved, so that discrete modes can emerge in the continuum. These types of modes are referred to as kinetic TAEs (KTAEs). Subsequent to the correction of gyrokinetic theory in [52], several missing FLR effects have been recovered. Consequently, KTAE theories definitely need to be reevaluated.

2.7 The kinetically driven MHD modes

In this section, we describe the kinetically driven modes (KDMs), such as the kinetic ballooning modes (KBMs), the energetic particle modes (EPMs), etc. In contrast to

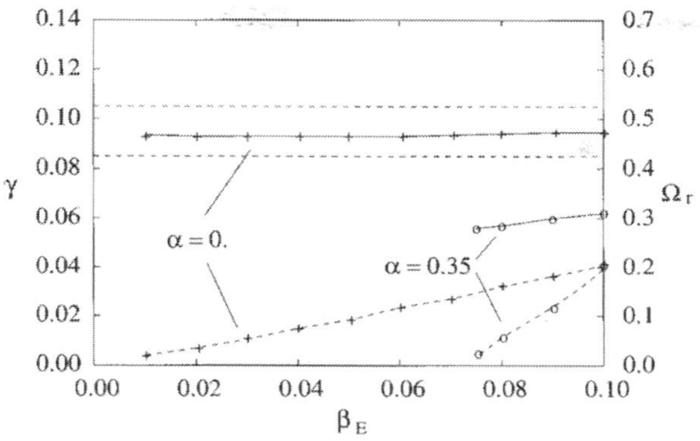

Figure 2.8. The growth rates (dashed curves) and real frequencies (solid curves) versus the energetic particle drive β_E for the TAE ($\alpha = 0$) and EPM cases ($\alpha = 0.35$). (Reproduced with permission from [54]. Copyright 2000, AIP Publishing LLC.)

the MHD eigenmodes discussed in the previous sections, the frequencies of these modes usually reside in the continuum spectrum. Therefore, they are generally damped without driving effects. Unlike the kinetic TAEs, for which the FLR effects are taken into account to resolve the singularity, for KDMs the strong kinetic effects from the wave–particle resonances are included to overcome continuum damping. This is why they are referred to as KDMs. The energetic particle drives to the marginal stable TAEs can instantly excite unstable TAEs, but the drives to the KDMs need to accumulate sufficient energy to overcome continuum damping for an unstable KDM to develop [53, 54]. Figure 2.8 shows the growth rates and real frequencies versus the energetic particle drive for TAE and EPM cases from a simulation. The two horizontal dashed lines indicate the TAE gap. For the $\alpha = 0$ case, the frequency falls inside the gap. It is therefore a TAE. The corresponding growth rate is directly proportional to the energetic beta β_E in this case. There is no driving delay. For the EPM case ($\alpha = 0.35$), however, the frequency resides in the lower continuum. The mode only starts to develop when a sufficient β_E (around 0.072) is applied. The missing energetic particle energy is absorbed by the continuum damping.

In sections 2.4 and 2.6 we have seen that there are two types of modes, ballooning and TAEs, according to the mode resonance features. The ballooning modes resonate at the rational surface. The TAEs, instead, resonate at the middle point between two rational surfaces with two Fourier harmonics coupling. Therefore, there are also two types of KDMs. Those related to the ballooning modes are referred to as KBMs, while the EPMs are related to the TAEs and are usually driven by the resonances between the waves and energetic particles. We employ the ballooning representation formalism to discuss them.

We start with the ballooning mode equation in the intermediate frequency regime, (2.92), with the energetic particle effects included. Introducing the transformation $\zeta = \delta\varphi p^{1/2}$, equation (2.92) becomes

$$\frac{\partial^2 \zeta}{\partial \theta^2} + \Omega^2(1 + 2\varepsilon \cos \theta)\zeta + \frac{\alpha \cos \theta}{p}\zeta - \frac{(s - \alpha \cos \theta)^2}{p^2}\zeta$$

$$- \frac{4\Gamma g^2}{p^2} + \frac{1}{p^{1/2}} \int \frac{\mathrm{d}\varepsilon \mathrm{d}\mu B}{|v_{\parallel}|} \omega_{\mathrm{d}} \delta g_{\mathrm{h}} = 0, \tag{2.106}$$

$$\mathbf{v}_{\parallel} \cdot \nabla \delta g_{\mathrm{h}} - \mathrm{i}\omega \delta g_{\mathrm{h}} = \mathrm{i}\omega \left(\mu B + v_{\parallel}{}^2\right) \frac{\partial F_{0\mathrm{h}}}{\partial \varepsilon} \left(\kappa_{\mathrm{r}} + \kappa_\theta \Lambda_s\right) p^{-1/2} \zeta,$$

where $p = 1 + \Lambda_s^2$, $g = \cos \theta + \Lambda \sin \theta$, $\Lambda_s = s(\theta - \theta_{\mathrm{k}}) - \alpha \sin \theta$, δg_{h} is the perturbed distribution function for hot ions, κ_{r} and κ_θ are, respectively, the radial and poloidal components of the magnetic field line curvature $\boldsymbol{\kappa}$, $\Omega = \omega/\omega_{\mathrm{A}}$, ω_{d} is the magnetic drift frequency, $\mathbf{v}$ is the particle speed, the subscripts $\perp$ and $\parallel$ represent, respectively, the perpendicular and parallel components to the equilibrium magnetic field line, $\epsilon = v^2/2$ is the particle energy, $\mu = v_\perp^2/2B$ is the magnetic moment and $F_{0\mathrm{h}}$ is the equilibrium distribution function for hot ions. For simplicity we have neglected the FLR effects and only take into account the kinetic effects from the energetic ions.

To study KDMs one need to investigate the singular layer behavior. In the ballooning representation space, the singular layer corresponds to the $\theta \to \infty$ limit. Again, we exclude the second TAE from discussion (i.e. we assume $\Gamma = 0$). Equation (2.106) in the $\theta \to \infty$ limit becomes:

$$\frac{\partial^2 \zeta}{\partial \theta^2} + \Omega^2(1 + 2\varepsilon \cos \theta)\zeta = 0. \tag{2.107}$$

This is the well-known Mathieu equation. According to Floquet's theorem, its solution takes following form

$$\zeta(\theta) = P(\theta) \exp\{\mathrm{i}\gamma\theta\},$$

where $P(\theta + 2\pi) = P(\theta)$. Since modes with longer parallel-to-**B** wavelengths tend to be more unstable, we shall examine the solutions corresponding to the two lowest periodicities. The first one is related to KBMs [53] and the second one is related to EPMs [54].

We first discuss KBMs. The KBM-type solution is given by

$$\zeta_{\mathrm{K}} = \exp\left\{\mathrm{i}\gamma_{\mathrm{K}}\theta\right\}\left(A_0 + A_2 \cos \theta + \cdots\right). \tag{2.108}$$

Inserting (2.107) into (2.106), one obtains, noting $\varepsilon \ll 1$,

$$\gamma_{\mathrm{K}}^2 \approx \Omega^2\left(1 + 2\varepsilon^2\Omega^2\right), \qquad \frac{A_2}{A_0} \approx 2\varepsilon\Omega^2.$$

Therefore, in the leading order, one has

$$\zeta_{\mathrm{K}} = \exp\{\mathrm{i}\Omega|\theta|\}, \tag{2.109}$$

where $\mathfrak{Im}\{\Omega\} > 0$ is imposed for causality. Note here that (2.109) is valid for general Ω, so that the frequency at the continuum is allowed as soon as the causality condition is satisfied.

Here, we should point out that, strictly speaking, the outer solution specified by (2.109) does not correspond to the ballooning modes, but to the internal kink modes. The internal kink mode has the same type of singular layer equation as that leading to (2.109) [43, 55]. The singular layer equation for KBMs was actually derived in [39]. Its structure is similar to the ideal MHD singular layer equation, leading to the well-known Mercier criterion [24, 26]. The kinetic effects modify the so-called apparent mass effects. Because of the wave–particle resonances, the apparent mass becomes frequency-dependent and complex. This changes the causality condition in the outer region ($\theta \to \infty$) and defines the actual KBMs.

Next, we discuss KDMs of TAE type, e.g. EPMs. These type of solutions can be expressed as [54]

$$\zeta_{\rm T} = \exp\left\{{\rm i}\gamma_{\rm T}\theta\right\}\left[A_1 \cos(\theta/2) + B_1 \sin(\theta/2) + \cdots\right]. \tag{2.110}$$

Inserting (2.110) into (2.107), one obtains, for $|\Omega^2 - 1/4| \sim O(\varepsilon)$ and

$$\gamma_{\rm T} = \left[\left(\Omega^2 - \Omega_+^2\right)\left(\Omega^2 - \Omega_-^2\right)\right]^{1/2}, \quad \frac{B_1}{A_1} = \left(\frac{\Omega^2 - \Omega_-^2}{\Omega_+^2 - \Omega^2}\right)^{1/2},$$

where $\Omega_\pm^2 = 1/4 \pm \varepsilon\Omega^2$. The leading order solution can, therefore, be expressed as

$$\zeta_{\rm T} = \exp\left\{{\rm i}\left[\left(\Omega^2 - \Omega_+^2\right)\left(\Omega^2 - \Omega_-^2\right)\right]^{1/2} |\theta|\right\}$$
$$\times\left[\cos(\theta/2) + \left(\frac{\Omega^2 - \Omega_-^2}{\Omega_+^2 - \Omega^2}\right)^{1/2} \sin(\theta/2)\right]. \tag{2.111}$$

The causality condition is $\mathfrak{Im}\{[(\Omega^2 - \Omega_+^2)(\Omega^2 - \Omega_-^2)]^{1/2}\} < 0$. Equation (2.111) can describe both TAEs and KDMs of TAE type (e.g. EPMs). The existence of a TAE solution requires the mode frequency to be in the gap, $\Omega_- < \Omega < \Omega_+$, as shown by the TAE theory in the configuration space in section 2.6. For KDMs the mode frequency can be in the continuum, i.e. outside the gap, as soon as the causality condition is satisfied. For TAEs $\zeta_{\rm T}$ contains an $O(1)$ back scattering and, hence, continuum damping is either suppressed or much reduced. On the other hand, for KDMs $\zeta_{\rm K}$ contains no back scattering from the periodic potential in (2.107) and, consequently, there is a significant amount of continuum damping. Note that, in principle, both types of solutions can co-exist at $|\Omega| \approx 1/2$. However, the TAE solution tends to be more unstable in this case than KDMs, since its continuum damping is much lower or absent.

With the outer solutions given by (2.109) or (2.111), one can obtain the corresponding dispersion relation by matching the outer and inner solutions. For

KBMs (2.106) can be used to construct the following quadratic form in the inner region:

$$2\,\zeta^{*}\frac{\mathrm{d}\zeta}{\mathrm{d}\theta}\bigg|_{-\infty}^{+\infty} + \delta W_{\mathrm{f}} + \delta W_{\mathrm{k}} = 0, \tag{2.112}$$

where

$$\delta W_{\mathrm{f}} = \int_{-\infty}^{+\infty} \mathrm{d}\theta \left\{ \left|\frac{\partial\zeta}{\partial\theta}\right|^2 - \left[\frac{\alpha\cos\theta}{p} - \frac{(s-\alpha\cos\theta)^2}{p^2}\right]|\zeta|^2 \right\},$$

$$\delta W_{\mathrm{k}} = \int_{-\infty}^{+\infty} \mathrm{d}\theta\zeta^{*}\frac{1}{p^{1/2}}\int\frac{\mathrm{d}\varepsilon\mathrm{d}\mu B}{|v_{\parallel}|}\omega_{\mathrm{d}}\delta g_{\mathrm{h}}.$$

Here, the superscript $*$ represents the complex conjugate. Matching the inner (2.112) and outer (2.109) solutions one obtains the dispersion relation [53]

$$-\mathrm{i}\Omega + \delta W_{\mathrm{f}} + \delta W_{\mathrm{k}} = 0. \tag{2.113}$$

Here, we note that the kinetic effects from the core plasma should also be taken into account in the outer region. As proved in [39], this results in the so-called apparent mass effect and causes Ω in the first term of (2.113) to become a complicated function of the actual mode frequency.

Similarly, for KDMs of TAE type, for example EPMs, one needs to consider the even and odd modes. For even modes the dispersion relation is given by [54] and [53]

$$-\mathrm{i}\left(\frac{\Omega_{-}^2-\Omega^2}{\Omega_{+}^2-\Omega^2}\right)^2 + \delta T_{\mathrm{f}} + \delta T_{\mathrm{k}} = 0, \tag{2.114}$$

where δT_{f} represents the MHD fluid contribution and δT_{k} is the energetic particle contribution to the quadratic form in the inner region.

In this section, we have discussed the KDMs. They are characterized by the modification of singular layer physics by wave–particle resonance effects. The resonance effects cause the eigenvalue problem to become a complex problem. The complex eigenvalue can change the causality conditions, so that new modes can exist in the domain where the ideal MHD modes are forbidden. The FLR effects can also change the singular layer equations, leading to new types of modes (see chapter 4), which are conventionally assigned different names, such as KTAEs, etc.

2.8 Discussion

In this chapter, we have given an overview of the ideal MHD theory in toroidally confined fusion plasmas. We first describe the MHD spectrum and global stability theory, and then four types of fundamental MHD modes, interchange, ballooning, TAEs and KDMs, are discussed. In describing these modes we detail the fundamental analytical treatments of MHD modes in toroidal geometry, such as the average technique for singular layer modes, the ballooning representation and the mode coupling treatment in the TAE/KDM theories, etc. Note that the analytical approach is often limited in toroidal plasma physics. The global numerical treatment of MHD modes is also reviewed in this chapter, especially the AEGIS code formalism. These theories are described basically in the ideal MHD framework. Here, we briefly discuss the kinetic and resistive modifications to the ideal MHD theory.

Let us first discuss the kinetic effects. Since there is a strong magnetic field in magnetically confined fusion plasmas, MHD theory can be quite good in describing plasma behaviors in the direction perpendicular to the magnetic field. This is because a strong magnetic field can hold plasmas together in perpendicular motion. Therefore, the MHD theory is a good model to describe perpendicular physics, if the FLR effects are insignificant. However, in the parallel direction the Lorentz force vanishes and particle collisions are insufficient to keep the particles moving collectively as a fluid cell. Consequently, a kinetic description in the parallel direction is generally necessary. The kinetic effect is particularly important when the wave–particle resonance effect prevails in the comparable frequency regime $\omega \sim \omega_{si}$ [39]. In the low frequency regime $\omega \ll \omega_{si}$, the wave–particle resonances can be so small that the kinetic description only results in an enhancement of the apparent mass effect. The kinetic effect in this case can be included by introducing an enhanced apparent mass. Another non-resonance case is the intermediate frequency regime $\omega_{si} \ll \omega \ll \omega_{se}$. In this regime the kinetic description results in a modification of the ratio of special heats. By introducing a proper ratio of specific heats the MHD model can still be a good approximation. Recovery of perpendicular MHD from gyrokinetics has been studied in detail in [52] and will be described in chapter 4.

Next, let us discuss the resistivity effects. Resistivity is usually small in magnetically confined fusion plasmas. Due to its smallness, resistivity effects are only important in the singular layers. The resistivity effects will be addressed in chapter 3. Here, we point out that, when the kinetic enhancement of the apparent mass effect is taken into account, the ratio of the resistivity and inertia layer widths changes. This leads the kinetic description of resistive MHD modes to become considerably different from the fluid description [56, 57]. A kinetic analysis of low frequency resistive MHD modes also becomes necessary.

The driving force for ideal MHD instabilities is related to the pressure gradient. Resistivity can instead cause field line reconnection and induce the so-called tearing modes. It is important to note that, if the current gradient is taken into account, the pressure driven modes and tearing modes can become coupled to each other. The underlying driving mechanism for pressure driven modes is the release of plasma thermal energy from the interchange of magnetic flux tubes. In fact, the interchange-type modes not only exchange thermal and magnetic energies between the flux tubes, but also the current. In a plasma with a current (or resistivity) gradient, such an interchange can create a current sheet at a mode resonance surface and result in the excitation of current interchange tearing modes [58]. These topics will also be discussed in chapter 3.

Bibliography

[1] Zheng L J 2012 Overview of magnetohydrodynamics theory in toroidal plasma confinement *Topics in Magnetohydrodynamics* ed L J Zheng (Rijeka: InTech) http://www.intechopen.com/books/topics-in-magnetohydrodynamics/overview-of-magnetohydrodynamics-theory-in-toroidal-plasma-confinement

[2] Zheng L-J and Tessarotto M 1994 Collisionless kinetic ballooning mode equation in the low-frequency regime *Phys. Plasmas* **1** 3928–35

[3] Zheng L-J and Chen L 1998 Plasma compressibility induced toroidal Alfvén eigenmode *Phys. Plasmas* **5** 444–9

[4] Bernstein I B, Frieman E A, Kruskal M D and Kulsrud R M 1958 An energy principle for hydromagnetic stability problems *Proc. R. Soc.* A **244** 17–40

[5] Frieman E and Rotenberg M 1960 On hydromagnetic stability of stationary equilibria *Rev. Mod. Phys.* **32** 898–902

[6] Zheng L-J, Chu M S and Chen L 1999 Effect of toroidal rotation on the localized modes in low beta circular tokamaks *Phys. Plasmas* **6** 1217–26

[7] Correa-Restrepo D 1982 Resistive ballooning modes in three-dimensional configurations *Z. Naturforsch.* A **37** 848

[8] Grimm R C, Greene J M and Johnson J L 1976 *Methods of Computational Physics* (New York: Academic)

[9] Chance M, Greene J, Grimm R, Johnson J, Manickam J, Kerner W, Berger D, Bernard L, Gruber R and Troyon F 1978 Comparative numerical studies of ideal magnetohydrodynamic instabilities *J. Comput. Phys.* **28** 1–13

[10] Bernard L, Helton F and Moore R 1981 Gato: an MHD stability code for axisymmetric plasmas with internal separatrices *Comput. Phys. Commun.* **24** 377

[11] Glasser A 1997 The direct criterion of Newcomb for the stability of an axisymmetric toroidal plasma *Los Alamos Report* LA-UR-95-528

[12] Zheng L-J and Kotschenreuther M 2006 AEGIS: An adaptive ideal-magnetohydrodynamics shooting code for axisymmetric plasma stability *J. Comput. Phys.* **211** 748–66

[13] Zheng L-J, Kotschenreuther M and Chu M S 2005 Rotational stabilization of resistive wall modes by the shear Alfvén resonance *Phys. Rev. Lett.* **95** 255003

[14] Zheng L J, Kotschenreuther M T and van Dam J W 2010 AEGIS-K code for linear kinetic analysis of toroidally axisymmetric plasma stability *J. Comput. Phys.* **229** 3605–22

[15] Cheng C Z and Chance M S 1986 Low-n shear Alfvén spectra in axisymmetric toroidal plasmas *Phys. Fluids* **29** 3695–701

[16] Turnbull A D, Strait E J, Heidbrink W W, Chu M S, Duong H H, Greene J M, Lao L L, Taylor T S and Thompson S J 1993 Global Alfvén modes: theory and experiment *Phys. Fluids* B **5** 2546–53

[17] Chen E, Berk H, Breizman B and Zheng L J 2010 Continuum damping of free-boundary TAE with AEGIS *Bull. Am. Phys. Soc.* **55** 277

[18] Zheng L J, Kotschenreuther M and Valanju P 2013 Behavior of $n = 1$ magnetohydrodynamics modes of infernal type at high-mode pedestal with plasma rotation *Phys. Plasmas* **20** 012501

[19] Degtyarev L, Martynov A, Medvedev S, Troyon F, Villard L and Gruber R 1997 The KINX ideal MHD stability code for axisymmetric plasmas with separatrix *Comput. Phys. Commun.* **103** 10–27

[20] Chance M S 1997 Vacuum calculations in azimuthally symmetric geometry *Phys. Plasmas* **4** 2161–80

[21] Mercier C 1962 Critere de stabilite d'an systeme toroidal hydromagnetique en pression scalaire *Nucl. Fusion Suppl.* **2** 801

[22] Greene J M and Johnson J L 1962 Stability criterion for arbitrary hydromagnetic equilibria *Phys. Fluids* **5** 510–7

[23] Glasser A H, Greene J M and Johnson J L 1975 Resistive instabilities in general toroidal plasma configurations *Phys. Fluids* **18** 875–88

[24] Greene J M and Johnson J L 1968 Interchange instabilities in ideal hydromagnetic theory *Plasma Phys.* **10** 729

[25] Johnson J L and Greene J M 1967 Resistive interchanges and the negative v'' criterion *Plasma Phys.* **9** 611

[26] Connor J W, Hastie R J and Taylor J B 1979 High mode number stability of an axisymmetric toroidal plasma *Proc. R. Soc.* A **365** 1–17

[27] Berk H L, Rosenbluth M N and Shohet J L 1983 Ballooning mode calculations in stellarators *Phys. Fluids* **26** 2616–20

[28] Mikhailovsky A 1974 *Theory of Plasma Instabilities* (New York: Consultants Bureau)

[29] Newcomb W A 1960 Hydromagnetic stability of a diffuse linear pinch *Ann. Phys. NY* **10** 232–67

[30] Zheng L-J and Tsai S T 1992 Energetic particle modified mercier criterion *Phys. Fluids* B **4** 1416–9

[31] Ware A A and Haas F A 1966 Stability of a circular toroidal plasma under average magnetic well conditions *Phys. Fluids* **9** 956–64

[32] Shafranov V D and Yurchenko E I 1967 *Zh. Eksp. Teor. Fiz.* **53** 1157 (in Russian) Shafranov V D and Yurchenko E I 1968 Condition for flute instability of a toroidal-geometry plasma *Sov. Phys.—JETP* **26** 682–6 (Engl. transl.)

[33] Chance M S *et al* 1979 MHD stability limits on high-beta tokamaks *Proc. 7th Int. Conf. Plasma Physics and Controlled Fusion Research, (Innsbruck, Austria, 23–30 August, 1978)* vol 1 (Vienna: International Atomic Energy Agency) p 677

[34] Zheng L J 2012 Free boundary ballooning mode representation *Phys. Plasmas* **19** 102506

[35] Lee Y C and van Dam J W 1977 Kinetic theory of ballooning instabilities, *Proc. Finite Beta Theory Workshop* vol CONF-7709167, ed B Coppi and W Sadowski (Washington, D.C.: US Department of Energy) p 93

[36] Hazeltine R D, Hitchcock D A and Mahajan S M 1981 Uniqueness and inversion of the ballooning representation *Phys. Fluids* **24** 180–1

[37] Connor J W, Hastie R J and Taylor J B 1978 Shear, periodicity, and plasma ballooning modes *Phys. Rev. Lett.* **40** 396–9

[38] Greene J and Chance M 1981 The second region of stability against ballooning modes *Nucl. Fusion* **21** 453

[39] Zheng L-J and Tessarotto M 1994 Collisionless kinetic ballooning equations in the comparable frequency regime *Phys. Plasmas* **1** 2956–62

[40] Lortz D and Nührenberg J 1978 Ballooning stability boundaries for the large-aspect-ratio tokamak *Phys. Lett.* A **68** 49–50

[41] Tang W, Connor J and Hastie R 1980 Kinetic-ballooning-mode theory in general geometry *Nucl. Fusion* **20** 1439

[42] Lortz D 1975 The general 'peeling' instability *Nucl. Fusion* **15** 49

[43] Wesson J 1978 Hydromagnetic stability of tokamaks *Nucl. Fusion* **18** 87

[44] Snyder P B, Wilson H R, Ferron J R, Lao L L, Leonard A W, Osborne T H, Turnbull A D, Mossessian D, Murakami M and Xu X Q 2002 Edge localized modes and the pedestal: a model based on coupled peeling–ballooning modes *Phys. Plasmas* **9** 2037–43

[45] Shi T, Wang H and Zheng L J 2013 private communication

[46] Wilson H R, Snyder P B, Huysmans G T A and Miller R L 2002 Numerical studies of edge localized instabilities in tokamaks *Phys. Plasmas* **9** 1277–86

[47] Cheng C, Chen L and Chance M 1985 High-*n* ideal and resistive shear Alfvén waves in tokamaks *Ann. Phys.* **161** 21–47

[48] Rosenbluth M N, Berk H L, Van Dam J W and Lindberg D M 1992 Mode structure and continuum damping of high-*n* toroidal Alfvén eigenmodes *Phys. Fluids* B **4** 2189–202

[49] Betti R and Freidberg J P 1992 Stability of Alfvén gap modes in burning plasmas *Phys. Fluids* B **4** 1465–74

[50] Vlad G, Zonca F and Briguglio S 1999 Dynamics of Alfvén waves in tokamaks *Nuovo Cimento* **22** 1–97

[51] Berk H L, Van Dam J W, Guo Z and Lindberg D M 1992 Continuum damping of low-*n* toroidicity-induced shear Alfvén eigenmodes *Phys. Fluids* B **4** 1806–35

[52] Zheng L J, Kotschenreuther M T and Van Dam J W 2007 Revisiting linear gyrokinetics to recover ideal magnetohydrodynamics and missing finite Larmor radius effects *Phys. Plasmas* **14** 072505

[53] Tsai S-T and Chen L 1993 Theory of kinetic ballooning modes excited by energetic particles in tokamaks *Phys. Fluids* B **5** 3284–90

[54] Zheng L-J, Chen L and Santoro R A 2000 Numerical simulations of toroidal Alfvén instabilities excited by trapped energetic ions *Phys. Plasmas* **7** 2469–76

[55] Chen L, White R B and Rosenbluth M N 1984 Excitation of internal kink modes by trapped energetic beam ions *Phys. Rev. Lett.* **52** 1122–5

[56] Zheng L-J and Tessarotto M 1996 Collisional effect on the magnetohydrodynamic modes of low frequency *Phys. Plasmas* **3** 1029–37

[57] Zheng L-J and Tessarotto M 1995 Collisional ballooning mode dispersion relation in the banana regime *Phys. Plasmas* **2** 3071–80

[58] Zheng L J and Furukawa M 2010 Current-interchange tearing modes: Conversion of interchange-type modes to tearing modes *Phys. Plasmas* **17** 052508

Chapter 3

Resistive MHD instabilities

In this chapter we describe the resistive MHD stability theory. We first discuss the resistive MHD theory for localized modes of Glasser, Greene and Johnson [1, 2]. Then the resistive ballooning mode theory is given and finally the tearing mode theory and discussion are presented.

Before describing the individual theories, we first give a schematic explanation of the resistive MHD modes, as compared to the ideal MHD modes. This is an extension to the introduction in [2]. In chapter 2 we discussed ideal MHD stability. In particular, we saw in section 2.2 that, if det $|\Xi_p|$ (or ξ for localized modes in section 2.3) has two zeros (or a small solution at the singular layer) inside the plasma, the system is unstable for internal modes. In figure 3.1 all the solid curves are the solutions of the ideal MHD equations with vanishing frequencies/growth rates and the dashed curves are the solutions of the resistive MHD equations. The curves E and E' depict the ideal MHD unstable cases: curve E has two zeros, while curve E' has one zero and is small at the singular layer. Therefore, according to Newcomb's theorem 5 [3], they are unstable. The solid curves F, F' and F'' are the ideal MHD stable cases, since they only cross the ψ axis once but are large at the singular layer.

Resistive MHD instabilities can still exist in ideal MHD stable cases. In figure 3.1 the dashed curves G and G', representing the resistive MHD solutions, connect the stable MHD modes across the singular layers to form unstable resistive MHD modes. There are two types of resistive MHD modes, the even and odd modes, according to the parities of the perturbed radial magnetic field δB_ψ. The even mode corresponds to the tearing modes, while the odd mode is related to the resistive interchange modes. Note that δB_ψ has the opposite parity to the radial displacement ξ. Therefore, in figure 3.1 the curve F–G–F' represents the resistive interchange modes, while the curve F–G'–F'' describes the tearing modes. For tearing modes there is a finite normal perturbed magnetic field at the singular layer. This causes the magnetic field lines to reconnect at the singular layer, so that magnetic islands and

doi:10.1088/978-1-6270-5423-2ch3

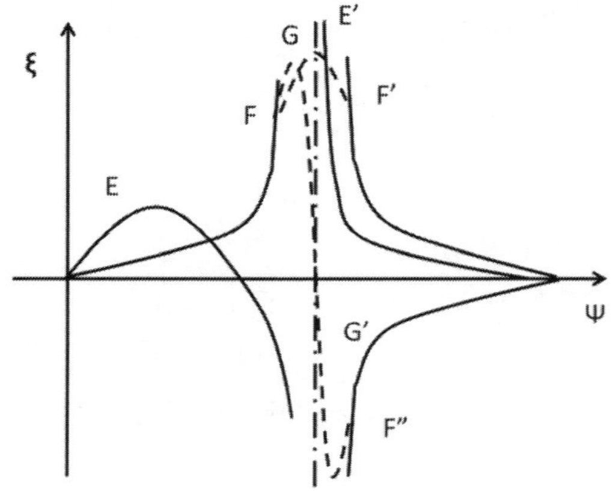

Figure 3.1. A schematic explanation of the ideal and resistive MHD instabilities. The dot-dashed line represents the singular layer.

stochastic field line regions emerge. Note that the interchange modes in tokamaks interchange not only the plasma and magnetic energies, but also the equilibrium current. This means that the interchange and tearing modes can become coupled.

In the ideal MHD theory one often uses Newcomb's theorem 5 to determine the stability, see the interchange and ballooning mode theories in sections 2.3 and 2.4. It is important to understand this issue from the eigenvalue perspective. Note that either ideal or resistive MHD eigenvalue problems are used to solve the governing differential equations under certain boundary conditions. The curves E and E' represent the ideal MHD stable cases, simply because (1) they are the solutions of the ideal MHD equations with a zero growth rate and (2) they can satisfy the boundary conditions at the left and right ends. The case E' represents the marginal stability case, as one cannot push the zero further toward the singular layer. In the case of curve E a marginal stability eigenmode solution exists between the origin and the point where the curve cuts the axis. However, when a growth rate is added, the cutting point can move further toward the singular layer. This indicates that modes with a finite growth rate can exist in the case E. Therefore, the case E is more unstable. From the eigenmode perspective the curves F, F' and F'' are classified as ideal MHD stable cases simply because one cannot construct a marginally stable solution that satisfies the boundary conditions on the other side. With the resistive solution connection across the singular layer, both the curves F–G–F' and F–G'–F'' become the solutions that satisfy the boundary conditions on both sides. The eigenvalue picture can apply to cases with a finite frequency. The same principle has been used in section 2.2 to discuss the global mode structure.

In the linear analyses we use the following resistive MHD equations:

$$\rho_m \gamma^2 \boldsymbol{\xi} = \delta \mathbf{J} \times \mathbf{B} + \mathbf{J} \times \delta \mathbf{B} - \nabla \delta P, \tag{3.1}$$

$$\delta \mathbf{B} + \frac{\eta}{\gamma} \nabla \times (\nabla \times \delta \mathbf{B}) = \nabla \times \boldsymbol{\xi} \times \mathbf{B}, \tag{3.2}$$

$$\mu_0 \delta \mathbf{J} = \nabla \times \delta \mathbf{B}, \tag{3.3}$$

$$\delta P = -\boldsymbol{\xi} \cdot \nabla P - \Gamma P \nabla \cdot \boldsymbol{\xi}. \tag{3.4}$$

The resistivity gradient effects will be included in a suitable place. The vorticity equation is still applicable for resistive MHD:

$$-\nabla \cdot \frac{\mathbf{B}}{B^2} \times \rho_m \gamma^2 \boldsymbol{\xi} = \mathbf{B} \cdot \nabla \frac{\mathbf{B} \cdot \delta \mathbf{J}}{B^2} + \delta \mathbf{B} \cdot \nabla \sigma - \mathbf{J} \cdot \nabla \frac{\mathbf{B} \cdot \delta \mathbf{B}}{B^2} + \nabla \times \frac{\mathbf{B}}{B^2} \cdot \nabla \delta P. \tag{3.5}$$

In the following sections, we will use this set of resistive MHD equations to study various resistive MHD modes.

3.1 The resistive MHD theory of Glasser, Greene and Johnson

In this section, we derive the resistive MHD singular layer equations of Glasser, Greene and Johnson [1, 2]. The derivation is an extension to the ideal MHD singular layer theory in section 2.3. The derivation is tedious and lengthy (we also refer partially to [4]). Therefore, let us first list the final resistive MHD singular layer equations and then describe their derivations one by one. The closed set of equations are as follows:

$$\Psi_{XX} - H\Gamma_X = Q(\Psi - X\Xi), \tag{3.6}$$

$$Q^2 \Xi_{XX} - QX^2 \Xi + E\Upsilon + QX\Psi + \Gamma = 0, \tag{3.7}$$

$$Q\Upsilon_{XX} - X^2\Upsilon - GQ^2\Upsilon + (G - KE)Q^2\Xi + X\Psi - KQ^2\Gamma = 0, \tag{3.8}$$

$$\Gamma_X = H\Psi_{XX} + F\Upsilon_X, \tag{3.9}$$

where E, F, H and M have been given in section 2.3, and

$$G = \frac{\langle B^2 \rangle}{M\Gamma P}, \qquad K = \frac{\Lambda^2}{MP'^2} \frac{\langle B^2 \rangle}{\langle B^2/|\nabla V|^2 \rangle},$$
$$Q_0 = \left(\frac{\eta \alpha^2 \Lambda^2 \langle B^2 \rangle}{\rho M \langle B^2/|\nabla V|^2 \rangle} \right)^{1/3}, \qquad Q = \frac{\gamma}{Q_0}, \tag{3.10}$$
$$X_0 = \left(\frac{\rho M \eta^2 \langle B^2 \rangle^2}{\alpha^2 \Lambda^2 \langle B^2/|\nabla V|^2 \rangle^2} \right)^{1/6}, \qquad X = \frac{V - V_0}{X_0}.$$

Here, the perturbed quantities are given as follows:

$$\Psi = \langle \delta\mathbf{B} \cdot \nabla V\rangle / \mathrm{i}\alpha\Lambda X_0,$$

$$\Gamma = \frac{\left\langle B^2/|\nabla V|^2 \right\rangle}{\Lambda^2}\frac{p'}{\chi'}\left[\langle \delta\mathbf{B} \cdot \nabla\theta\rangle - \Lambda\frac{\langle \mathbf{B} \cdot \nabla\theta \times \nabla V\rangle}{\left\langle B^2 \right\rangle}\Psi_X \right],$$

$$\Xi = \langle \boldsymbol{\xi} \cdot \nabla V\rangle, \qquad \Upsilon = \langle \mathbf{B} \cdot \delta\mathbf{B}\rangle / P'.$$

As in the ideal MHD case in section 2.3, the coordinates (x, u, θ) are employed. The perturbations can be assumed to vary as $\exp\{\mathrm{i}\alpha u\}$ with $\alpha = 2\pi n/\chi'$ as in section 2.3. Similarly, $\boldsymbol{\xi}$ and $\delta\mathbf{B}$ are projected in three directions as follows:

$$\boldsymbol{\xi} = \xi\frac{\nabla V}{|\nabla V|^2} + \mu\frac{\mathbf{B} \times \nabla V}{B^2} + \nu\frac{\mathbf{B}}{B^2}, \qquad \delta\mathbf{B} = b\frac{\nabla V}{|\nabla V|^2} + v\frac{\mathbf{B} \times \nabla V}{B^2} + \tau\frac{\mathbf{B}}{B^2}.$$

We consider only the singular layer modes whose wavelength across the magnetic surface $\lambda_\perp$ is much smaller than that on the surface and perpendicular to the magnetic field line $\lambda_\wedge$. The ordering assumptions are given as follows, in accordance with [1]:

$$\gamma \sim x \sim \epsilon, \qquad \frac{\partial}{\partial V} \sim \epsilon^{-1}, \qquad \frac{\partial}{\partial u} \sim \frac{\partial}{\partial \theta} \sim 1, \qquad \delta p \sim \epsilon, \qquad \eta \sim \epsilon^3,$$

and

$$\xi = \epsilon\xi^{(1)} + \cdots, \qquad \mu = \mu^{(0)} + \cdots, \qquad \nu^{(0)} = \nu^{(0)} + \cdots,$$
$$b = \epsilon^2 b^{(2)} + \cdots, \qquad v = \epsilon v^{(1)} + \cdots, \qquad \tau = \epsilon\tau^{(1)} + \cdots.$$

where $\epsilon \ll 1$ is the small parameter.

Some of the ideal MHD results are still valid here. Using the results in section 2.3 directly, one can reduce $\nabla \cdot \delta\mathbf{B} = 0$ to

$$\frac{\partial b^{(2)}}{\partial x} + \frac{1}{\Xi}\frac{\partial}{\partial u}v^{(1)} + \frac{J'}{P'}\frac{\partial v^{(1)}}{\partial \theta} - \frac{\chi'}{P'}\frac{\partial \sigma v^{(1)}}{\partial \theta} + \chi'\frac{\partial}{\partial \theta}\frac{\tau^{(1)}}{B^2} = 0. \tag{3.11}$$

After surface averaging it gives

$$\frac{\partial \bar{b}^{(2)}}{\partial x} + \frac{1}{\Xi}\frac{\partial \bar{v}^{(1)}}{\partial u} = 0. \tag{3.12}$$

Note also that the ∇V and ∇u projections of the induction equation (3.2) in the lowest order give, respectively,

$$\chi'\frac{\partial \xi^{(1)}}{\partial \theta} = 0, \qquad \chi'\frac{\partial \mu^{(0)}}{\partial \theta} = 0. \tag{3.13}$$

To satisfy the parallel component of (3.2) to the magnetic field line in the lowest order, one must require that

$$\left(\nabla \cdot \boldsymbol{\xi}_\perp\right)^{(0)} + 2\boldsymbol{\kappa} \cdot \boldsymbol{\xi}^{(0)} = \frac{\partial \xi^{(1)}}{\partial x} + \frac{1}{\Xi}\frac{\partial \mu^{(0)}}{\partial u} = 0. \tag{3.14}$$

where (2.43), (2.46) and (3.13) have been used.

Also, the two perpendicular components of the momentum equation (3.1) both give, in the lowest order,

$$\tau^{(1)} + \delta p^{(1)} = \tau^{(1)} - P'\xi^{(1)} - \Gamma P(\nabla\cdot\boldsymbol{\xi})^{(1)} = 0. \tag{3.15}$$

Using (2.60), one can prove that the oscillating part $\widetilde{\nabla\cdot\boldsymbol{\xi}}$ is of order ϵ^2. Therefore, only the average part $\langle\nabla\cdot\boldsymbol{\xi}\rangle$ is of order ϵ. Noting also (3.13), one can conclude from (3.15) that $\tau^{(1)}$ is independent of θ, as well as $\delta p^{(1)}$.

We now derive the first Glasser equation (3.6). With the ideal MHD results in mind we can focus on the resistivity effects. Noting that

$$\nabla V\cdot\nabla\times\nabla\times\delta\mathbf{B} = \frac{\partial}{\partial\theta}(\cdots) + \mathrm{i}\alpha\frac{|\nabla V|^2}{\Xi}\frac{\partial v^{(1)}}{\partial x},$$

the ∇V projection of the induction equation (3.2) in the next order gives

$$\bar{b}^{(2)} - \frac{\mathrm{i}\alpha\Lambda x}{\Xi}\xi^{(1)} = -\frac{\mathrm{i}\alpha}{\Xi}\frac{\eta}{q}\left\langle |\nabla V|^2\frac{\partial v^{(1)}}{\partial x}\right\rangle. \tag{3.16}$$

The vorticity equation (3.5) in the lowest order yields

$$\chi'\frac{\partial}{\partial\theta}\left(\frac{|\nabla V|^2}{B^2}\frac{\partial v^{(1)}}{\partial x}\right) - \frac{1}{P'}\chi'\frac{\partial\sigma}{\partial\theta}\frac{\partial\delta p}{\partial x} = 0.$$

Using (3.15), this equation is reduced to

$$\frac{\partial}{\partial\theta}\left(\frac{|\nabla V|^2}{B^2}\frac{\partial v^{(1)}}{\partial x}\right) + \frac{1}{P'}\frac{\partial\sigma}{\partial\theta}\frac{\partial\tau^{(1)}}{\partial x} = 0.$$

Integration over θ yields

$$\begin{aligned}\frac{\partial v^{(1)}}{\partial x} &= \frac{B^2/|\nabla V|^2}{\left\langle B^2/|\nabla V|^2\right\rangle}\frac{\partial\bar{v}^{(1)}}{\partial x} + \frac{1}{P'}\left(\frac{B^2/|\nabla V|^2}{\left\langle B^2/|\nabla V|^2\right\rangle}\left\langle\frac{B^2\sigma}{|\nabla V|^2}\right\rangle - \frac{B^2\sigma}{|\nabla V|^2}\right)\frac{\partial\tau^{(1)}}{\partial x}\\ &= -\frac{1}{\mathrm{i}\alpha}\frac{B^2/|\nabla V|^2}{\left\langle B^2/|\nabla V|^2\right\rangle}\frac{\partial^2\bar{b}^{(2)}}{\partial x^2} + \frac{1}{P'}\left(\frac{B^2/|\nabla V|^2}{\left\langle B^2/|\nabla V|^2\right\rangle}\left\langle\frac{B^2\sigma}{|\nabla V|^2}\right\rangle - \frac{B^2\sigma}{|\nabla V|^2}\right)\frac{\partial\tau^{(1)}}{\partial x},\end{aligned} \tag{3.17}$$

where (3.12) has been used.

Inserting (3.17) into (3.16) yields

$$\begin{aligned}&\bar{b}^{(2)} - \frac{\mathrm{i}\alpha\Lambda x}{\Xi}\xi^{(1)}\\ &= \frac{\eta}{q}\left\{\frac{\left\langle B^2\right\rangle}{\left\langle B^2/|\nabla V|^2\right\rangle}\frac{\partial^2\bar{b}^{(2)}}{\partial x^2} - \mathrm{i}\alpha\frac{1}{P'}\left(\frac{\left\langle B^2\right\rangle}{\left\langle B^2/|\nabla V|^2\right\rangle}\left\langle\frac{B^2\sigma}{|\nabla V|^2}\right\rangle - \left\langle B^2\sigma\right\rangle\right)\frac{\partial\tau^{(1)}}{\partial x}\right\}.\end{aligned} \tag{3.18}$$

Introducing the dimensionless parameters, one can find that this is the first Glasser equation (3.6).

To derive the second Glasser equation (3.7), we reduce the vorticity equation (3.5) as follows:

$$\frac{\rho q^2 |\nabla V|^2}{B^2}\frac{\partial \mu^{(0)}}{\partial x} = \frac{\partial}{\partial \theta}(\cdots)$$

$$+ \frac{J' - \sigma\chi'}{P'}\frac{\partial \sigma}{\partial \theta}\nu^{(1)} + \frac{\chi'}{B^2}\frac{\partial \sigma}{\partial \theta}\tau^{(1)} + \frac{\Lambda x}{\Xi}\frac{\partial}{\partial u}\left(\frac{|\nabla V|^2}{B^2}\frac{\partial \nu^{(1)}}{\partial x}\right) - \frac{P'}{\Xi}\frac{\partial}{\partial u}\left(\frac{\tau^{(1)}}{B^2}\right)$$

$$- \left(\frac{\nabla V \cdot \nabla\left(P + B^2\right)}{\Xi B^2 |\nabla V|^2} - \frac{\chi'}{P'}\Theta\frac{\partial \sigma}{\partial \theta}\right)\frac{\partial \tau^{(1)}}{\partial u} - \frac{\chi'}{P'}\frac{\partial \sigma}{\partial \theta}\frac{\partial}{\partial x}\left(\delta p^{(2)} - \xi^{(2)}P'\right) \tag{3.19}$$

Note that the term on the left can be reduced by (3.14) and, as shown in the derivation later on, the second and third terms on the right are canceled (see the first term on the right-hand side of (3.24)). We only need to work on the fourth term and the final term on the right.

We first simplify the fourth term on the right-hand side of (3.19). Using (3.17) and (3.18), one can prove that

$$\left\langle \frac{\Lambda x}{\Xi}\frac{\partial}{\partial u}\left(\frac{|\nabla V|^2}{B^2}\frac{\partial \nu^{(1)}}{\partial x}\right)\right\rangle$$

$$= -\frac{\Lambda x}{\Xi}\frac{1}{\left\langle B^2/|\nabla V|^2\right\rangle}\frac{\partial^2 \bar{b}^{(2)}}{\partial x^2} + \mathrm{i}\alpha\frac{\Lambda x}{\Xi}\frac{1}{P'}\left(\frac{1}{\left\langle B^2/|\nabla V|^2\right\rangle}\left\langle\frac{B^2\sigma}{|\nabla V|^2}\right\rangle - \langle\sigma\rangle\right)\frac{\partial \tau^{(1)}}{\partial x}$$

$$= -\frac{\Lambda x}{\Xi}\frac{1}{\left\langle B^2/|\nabla V|^2\right\rangle}\left\{\frac{q}{\eta}\frac{\left\langle B^2/|\nabla V|^2\right\rangle}{\left\langle B^2\right\rangle}\left(\bar{b}^{(2)} - \frac{\mathrm{i}\alpha\Lambda x}{\Xi}\xi^{(1)}\right)\right.$$

$$\left. + \mathrm{i}\alpha\frac{1}{P'}\left(\left\langle\frac{B^2\sigma}{|\nabla V|^2}\right\rangle - \frac{\left\langle B^2/|\nabla V|^2\right\rangle}{\left\langle B^2\right\rangle}\left\langle B^2\sigma\right\rangle\right)\frac{\partial \tau^{(1)}}{\partial x}\right\}$$

$$+ \mathrm{i}\alpha\frac{\Lambda x}{\Xi}\frac{1}{P'}\left(\frac{1}{\left\langle B^2/|\nabla V|^2\right\rangle}\left\langle\frac{B^2\sigma}{|\nabla V|^2}\right\rangle - \langle\sigma\rangle\right)\frac{\partial \tau^{(1)}}{\partial x}$$

$$= -\frac{q}{\eta}\frac{\Lambda x}{\Xi}\frac{1}{\left\langle B^2\right\rangle}\bar{b}^{(2)} + \mathrm{i}\alpha\frac{q}{\eta}\frac{\Lambda^2 x^2}{\Xi}\frac{1}{\left\langle B^2\right\rangle}\xi^{(1)} + \mathrm{i}\alpha\frac{\Lambda x}{\Xi}\frac{1}{P'}\left(\frac{\left\langle B^2\sigma\right\rangle}{\left\langle B^2\right\rangle} - \langle\sigma\rangle\right)\frac{\partial \tau^{(1)}}{\partial x}. \tag{3.20}$$

Next, we need to reduce the final term on the right of (3.19). This term represents the coupling of parallel motion. The equation of parallel motion reads

$$\rho_m \gamma^2 v^{(0)} = -P' b^{(2)} + \chi' \frac{\partial}{\partial \theta} \left(P' \xi^{(2)} - \delta p^{(2)} \right) + \frac{i\alpha \Lambda}{\Xi} x \tau^{(1)}. \tag{3.21}$$

From $\nabla \cdot \boldsymbol{\xi}^{(0)} = 0$ one can obtain

$$v^{(0)} = \left(\frac{B^2 \sigma}{P'} - \frac{B^2}{P'} \frac{\langle B^2 \sigma \rangle}{\langle B^2 \rangle} \right) \mu^{(0)} + \frac{B^2}{\langle B^2 \rangle} \bar{v}^{(0)}$$

$$= -\frac{1}{i\alpha} \left(\frac{B^2 \sigma}{P'} - \frac{B^2}{P'} \frac{\langle B^2 \sigma \rangle}{\langle B^2 \rangle} \right) \frac{\partial \xi^{(1)}}{\partial x} + \frac{1}{\rho_m \gamma^2} \frac{B^2}{\langle B^2 \rangle} \left(-P' b^{(2)} + \frac{i\alpha \Lambda}{\Xi} x \tau^{(1)} \right) \tag{3.22}$$

Here, $\bar{v}^{(0)}$ is obtained from (3.21) and $\mu^{(0)}$ from (3.14). Inserting this equation into (3.21), multiplying it by σ and taking the derivative with respective to x, one obtains

$$\frac{\chi'}{P'} \frac{\partial}{\partial x} \left\langle \left(\delta p^{(2)} - P' \xi^{(2)} \right) \frac{\partial \sigma}{\partial \theta} \right\rangle = - \left(\langle \sigma \rangle - \frac{\langle \sigma B^2 \rangle}{\langle B^2 \rangle} \right) \left[-\frac{\partial \bar{b}^{(2)}}{\partial x} + \frac{i\alpha \Lambda}{P' \Xi} \left(x \frac{\partial \tau^{(1)}}{\partial x} + \tau^{(1)} \right) \right]$$

$$+ \left\langle \sigma \frac{\partial \tilde{b}}{\partial x} \right\rangle - \frac{\rho q^2}{i\alpha} \frac{M_t}{\langle B^2 / |\nabla V|^2 \rangle} \frac{\partial^2 \xi^{(1)}}{\partial x^2}. \tag{3.23}$$

The second term on the right can be reduced as follows

$$\left\langle \sigma \frac{\partial \tilde{b}}{\partial x} \right\rangle = \left\langle \frac{J' - \sigma \chi'}{P'} \frac{\partial \sigma}{\partial \theta} v^{(1)} + \frac{\chi'}{B^2} \frac{\partial \sigma}{\partial \theta} \tau^{(1)} \right\rangle + i\alpha \langle \sigma \tilde{v}^{(1)} \rangle$$

$$= \left\langle \frac{J' - \sigma \chi'}{P'} \frac{\partial \sigma}{\partial \theta} v^{(1)} + \frac{\chi'}{B^2} \frac{\partial \sigma}{\partial \theta} \tau^{(1)} \right\rangle - \left(\frac{\langle B^2 \sigma / |\nabla V|^2 \rangle}{\langle B^2 / |\nabla V|^2 \rangle} - \langle \sigma \rangle \right) \frac{\partial \bar{b}^{(2)}}{\partial x}$$

$$+ \frac{i\alpha}{P'} \left(\frac{\langle B^2 \sigma / |\nabla V|^2 \rangle^2}{\langle B^2 / |\nabla V|^2 \rangle} - \left\langle \frac{B^2 \sigma^2}{|\nabla V|^2} \right\rangle \right) \tau^{(1)}. \tag{3.24}$$

Here, (3.11) has been used and, to reduce $\langle \sigma \tilde{v}^{(1)} \rangle$, equation (3.12) and the following equation have been used

$$\frac{1}{P'} \langle \sigma v^{(1)} \rangle = -\frac{1}{i\alpha P'} \frac{\langle \sigma B^2 / |\nabla V|^2 \rangle}{\langle B^2 / |\nabla V|^2 \rangle} \frac{\partial \bar{b}^{(2)}}{\partial x}$$

$$+ \frac{1}{P'^2} \left(\frac{\langle B^2 \sigma / |\nabla V|^2 \rangle^2}{\langle B^2 / |\nabla V|^2 \rangle} - \left\langle \frac{B^2 \sigma^2}{|\nabla V|^2} \right\rangle \right) \tau^{(1)}, \tag{3.25}$$

in which (3.17) has been used to express $v^{(1)}$.

Inserting (3.20) and (3.23) into (3.19) and averaging over θ, one obtains the second Glasser equation (3.7).

Next, we derive the third Glasser equation (3.8). The poloidal projection of induction equation (3.2) by the dot product with $\mathbf{B}\times\nabla V/|\nabla V|^2$ yields

$$\nu^{(1)} - \frac{\eta}{\gamma}|\nabla V|^2\frac{\partial^2 \nu^{(1)}}{\partial x^2} = \mathbf{B}\cdot\nabla\left(\frac{\boldsymbol{\xi}\cdot\mathbf{B}\times\nabla V}{|\nabla V|^2}\right) - \frac{(\mathbf{B}\times\nabla V)\cdot\nabla\times(\mathbf{B}\times\nabla V)}{|\nabla V|^4}\boldsymbol{\xi}\cdot\nabla V$$

$$= \chi'\frac{\partial\mu^{(1)}}{\partial\theta} + \mathrm{i}\alpha\Lambda x\mu^{(0)} - \frac{(\mathbf{B}\times\nabla V)\cdot\nabla\times(\mathbf{B}\times\nabla V)}{|\nabla V|^4}\xi^{(1)},$$

(3.26)

where (2.42) has been used. The parallel projection of the induction equation (3.2) by the dot product with $\mathbf{B}$ in the next order yields

$$\frac{1}{B^2}\tau^{(1)} - \frac{\eta}{\gamma}\frac{|\nabla V|^2}{B^2}\frac{\partial^2\tau^{(1)}}{\partial x^2} = \frac{P'}{B^2}\xi^{(1)} - (\nabla\cdot\boldsymbol{\xi})^{(1)} + \mathbf{B}\cdot\nabla\frac{\nu^{(1)}}{B^2} - 2(\boldsymbol{\xi}\cdot\boldsymbol{\kappa})^{(1)}$$

$$= \frac{P'}{B^2}\xi^{(1)} + \frac{P'}{\Gamma P}\xi^{(1)} - \frac{1}{\Gamma P}\tau^{(1)} + \chi'\frac{\partial}{\partial\theta}\frac{\nu^{(1)}}{B^2} + \frac{\Lambda x}{\Xi}\frac{\nu^{(0)}}{B^2}$$

$$- \xi^{(1)}\frac{\nabla V\cdot\nabla\left(2P + B^2\right)}{B^2|\nabla V|^2} - \frac{\chi'}{P'}\mu^{(1)}\frac{\partial\sigma}{\partial\theta}, \tag{3.27}$$

where (2.43), (2.40) and (3.15) have been used.

Using (3.26) to eliminate the $\frac{\chi'}{P'}\mu^{(1)}\frac{\partial\sigma}{\partial\theta}$ term in (3.27), one obtains

$$\left\langle\frac{1}{B^2}\right\rangle\tau^{(1)} - \frac{\eta}{\gamma}\left\langle\frac{|\nabla V|^2}{B^2}\right\rangle\frac{\partial^2\tau^{(1)}}{\partial x^2} - \frac{1}{P'}\left\langle\sigma\nu^{(1)}\right\rangle + \frac{\eta}{\gamma P'}\left\langle\sigma|\nabla V|^2\frac{\partial^2\nu^{(1)}}{\partial x^2}\right\rangle$$

$$= P'\left\langle\frac{1}{B^2}\right\rangle\xi^{(1)} + \frac{P'}{\Gamma P}\xi^{(1)} - \frac{1}{\Gamma P}\tau^{(1)} + \frac{\Lambda x}{\Xi}\left\langle\frac{\nu^{(0)}}{B^2}\right\rangle - \mathrm{i}\alpha\frac{1}{P'}\langle\sigma\rangle\Lambda x\mu^{(0)}$$

$$+ \left\langle-\frac{\nabla V\cdot\nabla\left(2P + B^2\right)}{B^2|\nabla V|^2} + \frac{\sigma}{P'}\frac{(\mathbf{B}\times\nabla V)\cdot\nabla\times(\mathbf{B}\times\nabla V)}{|\nabla V|^4}\right\rangle\xi^{(1)}. \tag{3.28}$$

Note that the last term on the right-hand side of this equation can be reduced by (2.60) in section 2.3, the third term on the left is given in (3.25) and the fourth term on the left can be reduced as follows

$$\frac{\eta}{qP'}\left\langle |\nabla V|^2\sigma\frac{\partial^2 v^{(1)}}{\partial x^2}\right\rangle = -\frac{\eta}{\mathrm{i}\alpha P'q}\frac{\left\langle \sigma B^2\right\rangle}{\left\langle B^2/|\nabla V|^2\right\rangle}\frac{\partial^3\bar{b}^{(2)}}{\partial x^3}$$
$$+\frac{\eta}{P'^2q}\left(\frac{\left\langle B^2\sigma/|\nabla V|^2\right\rangle\left\langle \sigma B^2\right\rangle}{\left\langle B^2/|\nabla V|^2\right\rangle} - \left\langle B^2\sigma^2\right\rangle\right)\frac{\partial^2\tau^{(1)}}{\partial x^2}.$$

Here, $\frac{\partial^3\bar{b}^{(2)}}{\partial x^3}$ can be reduced using (3.6). Therefore, the third Glasser equation (3.8) can be obtained from (3.28) after introducing the dimensionless parameters.

The fourth Glasser equation (3.9) just defines the unknown Γ for (3.6)–(3.8). Thus we complete the derivation of the four Glasser resistive MHD equations.

Now, let us discuss how to use the Glasser resistive MHD equations (3.6)–(3.8), with (3.9) used to eliminate the unknown Γ, to construct the resistive MHD stability problem [1]. Since the set (3.6)–(3.8) is of sixth order, one can expect six independent solutions. These resistive layer solutions need to be matched to the outer region solutions on both sides of the singular layer to construct the nonlocal stability problem. In the outer limit $X \to \infty$, four of the six solutions are exponentially increasing or decreasing as $\exp\{\pm X^2/2Q^{1/2}\}$. The increasing solutions are eliminated by the boundary conditions, while the decreasing solutions have no effect on the matching conditions. The two remaining solutions behave as powers of X as $X \to \pm\infty$:

$$\Psi \propto X^p, \qquad \Xi = \Upsilon = \Psi/X, \qquad p = 1/2 \pm \sqrt{-D_{\mathrm{I}}}, \tag{3.29}$$

where D_{I} is given in (2.65). These two solutions have the proper forms for matching to the outer solutions, which are governed by the ideal MHD. As discussed in section 2.3, the ideal MHD instabilities occur when $D_{\mathrm{I}} > 0$. If $D_{\mathrm{I}} < 0$, i.e. in the ideal MHD stable case, the ideal MHD outer solutions can be matched to the resistive MHD inner solutions with the asymptotic behavior given in (3.29) to obtain the dispersion relation for determining the resistive MHD stability.

From localized resistive MHD mode theory one can study various stability properties [1]. Alternatively, one can obtain the stability criteria from resistive ballooning mode theory. Since the ballooning mode framework is in essence a global framework through translational invariance, we will discuss this issue in the ballooning mode picture in the following section.

3.2 The resistive MHD ballooning modes

In this section, we describe the resistive ballooning mode theory [5, 6]. We first derive the resistive ballooning mode equations and then derive the singular layer equations from the asymptotic analyses of the ballooning equations. The stability criteria of resistive interchange modes, etc, are given at the end. We also discuss the kinetic effects on the resistive MHD theory, especially the small parallel ion velocity effects [7, 8].

In sections 2.3 and 3.1, we presented the singular layer theory. In the early ballooning mode theory it was known that the Mercier criterion can be alternatively obtained from the ballooning mode theory [9]. In this section, we describe this alternative singular layer theory for resistive MHD modes.

The ideal MHD ballooning mode theory is given in section 2.4. The ballooning mode representation described in section 2.4 will be used here as well. Briefly, we have $\nabla_\perp \gg \nabla_\parallel$, $\nabla_\perp = in\nabla\beta$, with $\beta = \zeta - q\theta$, and $\mathbf{B}\cdot\nabla = \chi'\partial/\partial\theta$. Two key resistive modifications need to be considered. First, the induction equation (3.2) becomes

$$\delta\mathbf{B} = \frac{\nabla\times\boldsymbol{\xi}\times\mathbf{B}}{1+\left(\eta^*/\gamma\right)|\nabla\beta|^2}, \tag{3.30}$$

where $\eta^* = n^2\eta$. Second, since the high-n modes are considered, one also has $\mathbf{B}\cdot\delta\mathbf{B} + \delta P = 0$. Using (3.30) and (2.43), this condition becomes

$$\begin{aligned} &-\left(B^2+\Gamma P\right)\nabla\cdot\boldsymbol{\xi} - B^2\cdot\nabla\frac{\mathbf{B}\cdot\boldsymbol{\xi}}{B^2} - 2B^2\boldsymbol{\kappa}\cdot\boldsymbol{\xi} \\ &-\frac{\eta^*}{\gamma}|\nabla\beta|^2\,\boldsymbol{\xi}\cdot\nabla P - \frac{\eta^*}{\gamma}|\nabla\beta|^2\,\nabla\cdot\boldsymbol{\xi} = 0. \end{aligned} \tag{3.31}$$

As in the ideal MHD ballooning mode case in section 2.4, according to (3.31), we can introduce the stream function $\delta\varphi$, such that $\boldsymbol{\xi}_\perp = \mathbf{B}\times\nabla\delta\varphi/B^2$. We also define $D = \Gamma P\nabla\cdot\boldsymbol{\xi}$. The resistive ballooning equations can be constructed from the vorticity equation (3.5) and the parallel projection of the momentum equation (3.1) as follows [6]:

$$\begin{aligned} &\frac{\partial}{\partial\theta}\left(\frac{|\nabla\beta|^2}{1+\left(\eta^*/\gamma\right)|\nabla\beta|^2}\frac{1}{B^2}\frac{\partial}{\partial\theta}\delta\varphi\right) + \frac{2P'}{\chi'^4}\left(\kappa_n + q\theta\kappa_g\right)\delta\varphi - \frac{\rho_m\gamma^2}{\chi'^2B^2}|\nabla\beta|^2\delta\varphi \\ &= -\frac{2P'}{\chi'^4}\left(\kappa_n+q\theta\kappa_g\right)D, \end{aligned} \tag{3.32}$$

$$\begin{aligned} &\frac{\partial}{\partial\theta}\left(\frac{1}{B^2}\frac{\partial D}{\partial\theta}\right) - \frac{\rho_m\gamma^2}{\chi'^2}\frac{B^2+\Gamma P}{B^2\Gamma P}D - \frac{\rho_m\eta^*\gamma}{\chi'^2B^2}|\nabla\beta|^2 D - \frac{2P'}{\chi'^4}\frac{\eta^*}{\gamma}\left(\kappa_n+q\theta\kappa_g\right)D \\ &= \frac{2\rho_m}{P'\chi'^2}\left(\frac{P'^2\eta^*}{\chi'^2\gamma\rho_m}+\gamma^2\right)\left(\kappa_n+q\theta\kappa_g\right)\delta\varphi, \end{aligned} \tag{3.33}$$

where $\kappa_n = \boldsymbol{\kappa}\cdot\nabla\theta\times\nabla\zeta$ and $\kappa_g = \boldsymbol{\kappa}\cdot\nabla v\times\nabla\theta = -(\chi'/2P')\mathbf{B}\cdot\nabla\sigma$. Note also that $|\nabla\beta|^2 = |\nabla\zeta|^2 - 2q'\theta\nabla V\cdot\nabla\zeta + q'^2\theta^2|\nabla V|^2$.

One can solve (3.32) and (3.33) using asymptotic analyses. We note that the resistivity effects are only important in the singular layer. In the ballooning mode representation space this corresponds to $\theta \sim 1/\epsilon_\theta \to \infty$ and is therefore referred to as the outer region. To proceed, we need to analyze the orderings of the resistive ballooning equations in the outer region. We consider the most general case in which $(\eta^*/\gamma)|\nabla\beta|^2 \sim 1$. This maintains the two terms in the denominator of the first term in (3.32) as the same order. We also limit ourselves to considering the low frequency regime. These considerations lead us to introduce the following optimal ordering scheme:

$$a^2\eta^*/\epsilon_\theta^2 \sim \omega, \qquad \epsilon_\theta \sim \beta_t^{1/2}(\omega/\omega_{si}), \qquad \omega \ll \omega_{si}. \tag{3.34}$$

Here, the minor radius a is introduced for the purposes of dimensionless analyses, β_t is the plasma beta, the ion acoustic frequency $\omega_{si} \sim v_{thi}/Rq$, with q assumed to be of order unity and v_{thi} the ion thermal velocity. Since the beta is trivially small, we assume in addition that

$$\beta_t \lesssim a/R. \tag{3.35}$$

We first examine the sound wave equation (3.33). Using (3.34), the orderings of four terms on the left-hand side of (3.33) can be obtained as follows:

$$\omega_{si}^2, \qquad \omega^2, \qquad \omega^2\beta_t, \qquad \omega_{si}\omega\beta_t^{1/2}, \tag{3.36}$$

while the two terms in the first bracket on the right-hand side can be estimated as follows

$$\frac{\omega^2}{p'^2\eta^*/\omega\rho_m} \sim \frac{a/R}{\beta_t}. \tag{3.37}$$

Using (3.34), we can also estimate the orderings for the first, second and final term in (3.32), respectively, as follows

$$\frac{1}{\epsilon_\theta^2}, \qquad \beta_t\frac{R}{a}\frac{1}{\epsilon_\theta}, \qquad \beta_t\frac{\omega^2}{\omega_{si}^2}\frac{1}{\epsilon_\theta^2}. \tag{3.38}$$

The term on the right of (3.32), which gives the so-called apparent mass effects, can be proved *a posteriori* to be the same order as the last term on the left.

Now, we consider the ideal MHD region. In this region the resistivity and inertia effects can be neglected and therefore one has

$$\frac{\partial}{\partial\theta}\left(\frac{|\nabla\beta|^2}{1+(\eta^*/\gamma)|\nabla\beta|^2}\frac{1}{B^2}\frac{\partial}{\partial\theta}\delta\varphi\right) + \frac{2P'}{\chi'^4}\left(\kappa_n + q\theta\kappa_g\right)\delta\varphi$$
$$= -\frac{2P'}{\chi'^4}\left(\kappa_n + q\theta\kappa_g\right)D, \tag{3.39}$$

$$\frac{\partial}{\partial\theta}\left(\frac{1}{B^2}\frac{\partial D}{\partial\theta}\right) = 0. \tag{3.40}$$

From (3.40) one obtains $D = 0$. In this case (3.39) becomes a second order equation and therefore has two independent solutions. From the asymptotic analyses for Mercier's criterion in section 2.3 we know that at the limit $\theta \to \infty$ the two independent solutions behave as the small and large solutions, so that the general solution can be expressed as follows

$$\delta\varphi = a_1 |\theta|^s + a_2 |\theta|^{-1-s}, \tag{3.41}$$

where a_1 and a_2 are constants and $s = -1/2 + \sqrt{-D_{\rm I}}$. We introduce the index for resistive interchange mode

$$D_{\rm R} = F + E + H^2. \tag{3.42}$$

Therefore, $-D_{\rm I} = 1/4 + H^2 - H - D_{\rm R}$. Here, the definitions of the parameters F, E and H are the same as those in the Mercier criterion in section 2.3.

Next, we consider the outer region. We introduce two length scales (θ, $\epsilon_\theta z$) and expand the unknowns as $\delta\varphi = \delta\varphi_0 + \epsilon_\theta \delta\varphi_1 + \cdots$ and $D = D_0 + \epsilon_\theta D_1 + \cdots$, where z corresponds to the slowly varying scale. We first solve (3.33). The lowest order equation reads:

$$\frac{\partial}{\partial\theta}\left(\frac{1}{B^2}\frac{\partial D_0}{\partial\theta}\right) = 0.$$

The solution of this equation is simply $D_0 = D_0(z)$. The next order equation is as follows

$$\begin{aligned}\frac{\partial}{\partial\theta}\left[\frac{1}{B^2}\left(\frac{\partial D_1}{\partial\theta} + \frac{\partial D_0}{\partial z}\right)\right] + \rho\omega^2\frac{1}{\gamma P}D_0 - \frac{2P'}{\chi'^4}\frac{\eta^*}{\gamma}q\theta\kappa_g D_0 \\ = \frac{2\rho_m}{P'\chi'^2}\left(\frac{P'^2\eta^*}{\chi'^2\gamma\rho_m} + \gamma^2\right)q\theta\kappa_g\delta\varphi_0.\end{aligned} \tag{3.43}$$

Keeping the second term in the equation of this order is based on the consideration that it is of order $(\omega_{\rm si}/\omega)\beta_{\rm t}^{1/2}$, which is much larger than ϵ_θ, as compared to the rest of the terms on the left-hand side. Noting that $\delta\varphi_0$ in the lowest order can be proved *a posteriori* to be independent of θ and $\kappa_g \propto \partial\sigma/\partial\theta$, the field line average ($\langle a\rangle \equiv \oint JB{\rm d}\theta a / \oint JB{\rm d}\theta$) of (3.43) yields $D_0 = 0$. Therefore, the solution of (3.43) can be obtained as

$$\frac{\partial D_1}{\partial\theta} = \frac{\rho_m}{P'^2}\left(\frac{P'^2\eta^*}{\chi'^2\gamma\rho_m} + \gamma^2\right)\left(B^2\sigma - \frac{B^2\langle B^2\sigma\rangle}{\langle B^2\rangle}\right)qz\delta\varphi_0. \tag{3.44}$$

We now move on to solve (3.32). In the lowest order one can obtain

$$\delta\varphi_0 = \delta\varphi_0(z). \tag{3.45}$$

Using (3.45), in the first order (3.32) can be solved

$$\delta\varphi_1(\theta, z) = f_1(z) + \frac{\mathrm{d}\delta\varphi_0/\mathrm{d}z}{\left\langle B^2/|\nabla V|^2\right\rangle + \left(\eta^*/\gamma\right)\left\langle B^2\right\rangle q'^2 z^2} \int\left[\left(\widetilde{\frac{B^2}{|\nabla V|^2}}\right) + \frac{\eta^*}{\gamma} q'^2 z^2 \widetilde{B^2}\right]\mathrm{d}\theta$$

$$+ \frac{\delta\varphi_0}{\chi'^2 q' z}\int\left[\left(\widetilde{\frac{B^2}{|\nabla V|^2}}\right) + \left(\widetilde{\frac{B^2\sigma}{|\nabla V|^2}}\right) + \frac{\eta^*}{\gamma}\left(\widetilde{B^2} + \widetilde{B^2\sigma z^2}\right)\right]\mathrm{d}\theta, \tag{3.46}$$

where $\tilde{a} = a - \langle a\rangle$. In the next order (3.32) gives the solubility condition for $\delta\varphi_2$:

$$\frac{\mathrm{d}}{\mathrm{d}z}\frac{x^2}{1+x^2}\frac{\mathrm{d}\delta\varphi_0}{\mathrm{d}x} + \frac{H(1-H)}{\left(1+x^2\right)^2}\delta\varphi_0 - \frac{H(1+H)x^2}{\left(1+x^2\right)^2}\delta\varphi_0 + D_{\mathrm{R}}\delta\varphi_0 - Q^3x^2\delta\varphi_0 = 0. \tag{3.47}$$

Here, equations (3.45)–(3.44) have been used, Q is defined in (3.10) in section 3.1 with the replacement

$$M_{\mathrm{t}} \to \left(1 + \frac{P'^2\eta^*}{\gamma^3\chi'^2\rho_m}\right)M_{\mathrm{t}} \tag{3.48}$$

for the parallel inertia parameter M_{t} inside the definition of M, and $x = [\eta^* q'^2 \langle B^2\rangle/\gamma\langle B^2/|\nabla V|^2\rangle]^{1/2} z$.

We need to solve (3.47) to obtain the solution for the resistive or inertia region. Letting

$$\delta\varphi_0 = |x|^s \exp\left\{(1+s-H)\frac{x^2}{2}\right\}\frac{\mathrm{d}^2}{\mathrm{d}x^2}\left[\exp\left\{\left(H-1-s-Q^{3/2}\right)\frac{x^2}{2}\right\}P(r)\right], \tag{3.49}$$

where $r = Q^{3/2}x^2$, one can transform (3.47) to the Kummer equation [10]:

$$r\frac{\mathrm{d}^2P}{\mathrm{d}r^2} + \left(\frac{1}{2} + s - r\right)\frac{\mathrm{d}P}{\mathrm{d}r} - \frac{1}{4}\left(Q^{3/2} + 1 + 2s - \frac{D_{\mathrm{R}}}{Q^{3/2}}\right)P = 0. \tag{3.50}$$

The solution, which remains finite and thus satisfies the boundary condition at ∞, is

$$P(r) \propto \frac{\pi}{\sin\pi\nu}\left[\frac{{}_1F_1\left(a^*, \nu, r\right)}{\Gamma\left(1 + a^* - \nu\right)\Gamma(\nu)} - r^{1-\nu}\frac{{}_1F_1\left(1 + a^* - \nu, 2 - \nu, r\right)}{\Gamma\left(a^*\right)\Gamma(2-\nu)}\right], \tag{3.51}$$

where $\nu = 1/2 + s$, $a^* = (Q^{3/2} + 2\nu - D_{\mathrm{R}}/Q^{3/2})/4$, Γ is the gamma function and ${}_1F_1$ is the Kummer function

$${}_1F_1\left(a^*, \nu, r\right) = 1 + \frac{a^*}{\nu}r + \frac{a^*\left(a^*+1\right)}{\nu(\nu+1)}\frac{r^2}{2!} + \cdots.$$

Therefore, the solution in (3.49) can be explicitly expressed as

$$\delta\varphi_0 = \frac{\pi|x|^s \exp\{-Q^{3/2}z^2/2\}}{\sin\pi\nu\Gamma(1+a^*-\nu)\Gamma(\nu)}\left\{\frac{Q^{3/2}-1-s+H}{2}\,{}_1F_1(a^*,\nu,r)\right.$$

$$+\frac{a^*-\nu}{\nu}\,{}_1F_1(a^*,\nu+1,r)-\frac{\Gamma(\nu)\Gamma(1+a^*-\nu)}{\Gamma(1-\nu)\Gamma(a^*)}Q^{\frac{3}{2}(\frac{1}{2}-s)}\,|x|^{-1-2s}$$

$$\left.\times\left[{}_1F_{1'}(a^*-\nu,1-\nu,r)+\frac{Q^{3/2}-1-s+H}{2(1-\nu)}x^2{}_1F_1(1+a^*-\nu,2-\nu,r)\right]\right\}. \tag{3.52}$$

In order to construct a solution that is valid both in the ideal MHD and resistive/inertia regions, we match the ideal solution in (3.41) for $\theta \to \infty$ to the resistive solution in (3.52) for $x \to 0$. For the matching, we note that the coordinates in the two regions are related as follows: $x^2 = (1/\theta_0^2 Q)\theta$, where $\theta_0^2 = \langle B^2/|\nabla V|^2\rangle Q_0/q'^2\langle B^2\rangle\eta^*$, where Q_0 is given in (3.10). We need to match the ratio of the large to small solutions between the two regions. In the ideal MHD region the ratio is, from (3.41),

$$\Delta' = a_2/a_1. \tag{3.53}$$

From (3.52) one can also find the ratio in the small x limit:

$$\Delta = \frac{4y_0^{1+2s}Q^{(5-2s)/4}}{Q^3-(1+s-H)^2}\frac{\Gamma(1/2+s)}{\Gamma(-1/2-s)}\frac{\Gamma\left[\left(Q^{3/2}+3-2s-D_R/Q^{3/2}\right)/4\right]}{\Gamma\left[\left(Q^{3/2}+1+2s-D_R/Q^{3/2}\right)/4\right]}. \tag{3.54}$$

Therefore, using (3.53) and (3.54), one can determine the matching condition or dispersion relation:

$$\Delta = \Delta'. \tag{3.55}$$

The dispersion relation, (3.55), can be used to study the resistive MHD stability condition. In particular, we note that, when $D_R > 0$, the gamma functions in (3.54) have an infinite sequence of poles due to the term $-D_R/4Q^{3/2}$. The resistive parameter Δ vanishes and diverges alternately, passing through all values many times. There are infinitely many roots Q for the dispersion relation, (3.55), for any given Δ' in the ideal MHD region. This indicates that there are resistive instabilities. Since y_0 scales as $\eta^{*-1/3}$, the factor y_0^{1+2s} in (3.55) is large. Consequently, the roots of (3.55) are in general near the poles of $\Gamma[(Q^{3/2}+3-2s-D_R/Q^{3/2})/4]$, that is

$$Q_k = -(1/2+s-2k)+\left[(1/2+s+2k)^2+D_R\right]^{1/2},$$

where k labels the root sequence. Noting that $\gamma \sim \eta^{*1/3}Q$ in the incompressible limit, i.e. $D=0$, the actual growth rate γ scales as $\eta^{*1/3}$. These kind of instabilities correspond to the well-known resistive interchange modes [1]. Due to the compressibility modification in (3.48) the actual dependence of γ on η^* can be more complicated. Inclusion of the compressibility effects causes the current results to be different from those in [6].

Here, we point out that the resistivity contributions to the final singular layer equation have two origins: one is from the resistive effect on the bending term (i.e. the first one in (3.32)); The other is from the resistive effect on the equation of parallel motion (i.e. from D_1). This gives rise to the resistive contribution to the apparent mass term (see (3.48) and the discussion).

Using the dispersion relation in (3.55) one can further study the case with $D_R \leqslant 0$. The analyses are similar to the case $D_R > 0$, including the linear tearing modes. Due to limitations of space the readers are directed to [1] and [6] for details.

3.3 The tearing mode and its coupling to interchange-type modes

In this section, we describe the nonlinear tearing mode theory. Unlike the interchange-type modes, in the tearing mode case the perturbed radial magnetic field is excited at the rational surface. This causes the magnetic field to reconnect to form magnetic islands and stochastic field line regions. This is a highly nonlinear problem. In this section we describe Rutherford's theory [11], with the inclusion of current interchange effects [12]. Current gradient effects can lead to coupling between the tearing and interchange modes.

We consider the slab model as [11]. The coordinate system (x, y, z) is shown in figure 3.2, in which x represents the distance from the rational surface, which labels the unperturbed magnetic surface and y is the binormal direction to the guiding magnetic field, which is in the z-direction.

Without losing generality, we can subtract the guiding field and use the relative field to study the tearing modes. Therefore, the equilibrium magnetic field can be expressed as $\mathbf{B} = (0, B_y' x, 0)$. Here, the prime denotes the derivative with respect to x and finite magnetic shear is assumed.

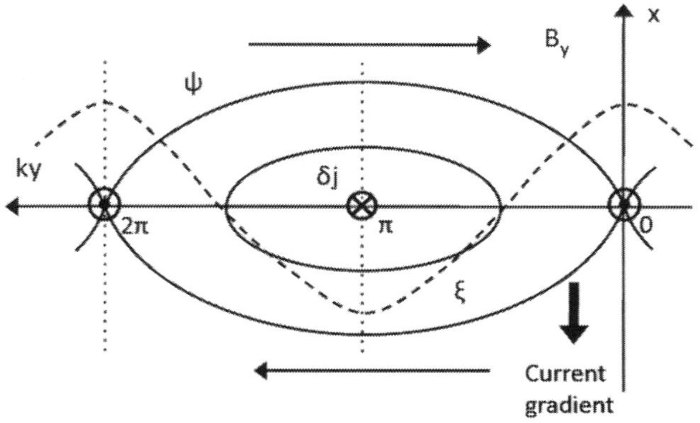

Figure 3.2. The coordinate system for analyzing the tearing modes.

Due to the divergence-free constraint of the magnetic-field $\nabla \cdot \mathbf{B} = 0$, one can introduce the magnetic stream function ψ to represent the magnetic field $B_x = -\partial\psi/\partial y$, $B_y = \partial\psi/\partial x$. Also, due to the suppression of the compressional Alfvén wave $\nabla \cdot \mathbf{v}_\perp = 0$ (see (3.14)), the velocity stream function φ can be introduced to represent the velocity $v_x = -\partial\varphi/\partial y$, $v_y = \partial\varphi/\partial x$. Here, the subscripts (x, y, z) are introduced to denote the corresponding projections.

The total magnetic flux can be written as

$$\psi(x, y, t) = \psi_0(x) + \delta\psi(y, t), \tag{3.56}$$

where $\psi_0(x) = B_y' x^2/2$ is the equilibrium value and $\delta\psi(y, t)$ is the perturbed value. We use the subscript 0 to denote the unperturbed quantities and δ to tag the perturbed quantities. The subscript 0 is dropped as soon as there is no ambiguity with the total quantities. The Fourier decomposition is introduced for the poloidal direction, such that $\delta\psi(y, t) = \delta\psi_1(t)\cos ky$, where k is the poloidal wave number. It is sufficient for our current focus to only keep the first harmonic.

To understand the derivation of the nonlinear tearing mode equation one needs to become familiar with the two coordinate systems: the total flux and Cartesian coordinates. From (3.56) one can obtain the transform from the total flux coordinate ψ to the radial Cartesian coordinate

$$x = \pm\sqrt{\frac{2}{B_y'}}\sqrt{\psi - \delta\psi}.$$

Defining the normalized flux coordinate $w = \psi/\delta\psi_1$ and $x_\mathrm{T} = 2\sqrt{\delta\psi_1/B_y'}$, one has

$$x = \pm\frac{x_\mathrm{T}}{2}\sqrt{w - \cos ky}. \tag{3.57}$$

From this equation we can discuss the island coordinate structure. The island 'O' point is located at $w = -1$ and $ky = \pi$, and the two 'X' points are at $w = 1$ and $ky = 0, 2\pi$, respectively, as shown in figure 3.2. Inside the magnetic island, one has $-1 \leqslant w \leqslant 1$ and the ky range is limited by the condition $w - \cos ky \geqslant 0$. Outside the magnetic island, one has the w range from 1 to $+\infty$ and $0 \leqslant ky \leqslant 2\pi$. The half island width can be obtained at $w = 1$ and $ky = \pi$, which is x_T.

We use Ampere's law and Ohm's law

$$\mu_0 \mathbf{J} = \nabla \times \mathbf{B}, \tag{3.58}$$

$$\eta\mathbf{J} = \mathbf{E} + \mathbf{v} \times \mathbf{B} \tag{3.59}$$

to construct the basic set of equations. Noting that the radial derivative with respect to x is dominant, the Ampere law in (3.58) gives

$$\frac{\mathrm{d}^2\delta\psi}{\mathrm{d}x^2} = \mu_0\delta j_z, \tag{3.60}$$

Noting that $\partial\mathbf{B}/\partial t = \nabla \times \mathbf{E}$, the application of the curl operation on Ohm's law in (3.59) and the projection of the resulting equation on the x-direction yield

$$\frac{\partial\delta\psi}{\partial t} - \left.\frac{\partial\delta\varphi}{\partial y}\right|_{\psi} B_y' x = \eta\delta j_z + \delta\eta j_z. \tag{3.61}$$

where we have noted that $\nabla x \cdot \nabla \times \mathbf{v} \times \mathbf{B} = \mathbf{B} \cdot \nabla v_x$, so that the derivative on $\delta\varphi$ in the second term is on the constant ψ. Here, the last term on the right-hand side represents the convective current interchange effect, with $\delta\eta = -\xi_x(\mathrm{d}\eta/\mathrm{d}x)$ [12]. The interchange-type modes exchange not only thermal and magnetic energies between the flux tubes, but also current. In a plasma with a resistivity (or current) gradient, such an interchange can create a current sheet at a mode resonance surface, as shown in figure 3.2. The final term in (3.61) describes exactly this effect. It is assumed that there is an interchange-type mode, $\xi_x = \xi_1 \cos ky$, in the system.

We perform the total flux surface average. Dividing by x and averaging over y at constant ψ to eliminate the second term on the left, equation (3.61) becomes

$$\frac{1}{\mu_0}\frac{\partial^2\delta\psi}{\partial x^2} = \frac{\left\langle \partial\delta\psi/\partial t/(\psi - \delta\psi)^{1/2}\right\rangle}{\eta\left\langle 1/(\psi - \delta\psi)^{1/2}\right\rangle} + \frac{\mathrm{d}\ln\eta}{\mathrm{d}x} j_z \frac{\left\langle \xi_x/(\psi - \delta\psi)^{1/2}\right\rangle}{\left\langle 1/(\psi - \delta\psi)^{1/2}\right\rangle}, \tag{3.62}$$

where $\langle\cdots\rangle = (k/2\pi)\int_0^{2\pi/k}\{\cdots\}\mathrm{d}y$ with ψ or w remaining constant. Here, we have used (3.60) to express δj_z on the left-hand side and noted that $\delta j_z(\psi)$ is a function of ψ only as required by the reduced vorticity equation $\mathbf{B}\cdot\nabla\delta j_z = 0$ [11]. Further integrating over x from $-\infty \to +\infty$, equation (3.62) becomes

$$\left.\frac{\partial\delta\psi}{\partial x}\right|_{-\infty}^{\infty} = -\frac{\mu_0}{\eta\sqrt{2B_y'}}\int_{-\infty}^{\infty}\frac{\mathrm{d}\psi}{(\psi - \delta\psi)^{1/2}}\frac{\left\langle \partial\delta\psi/\partial t/(\psi - \delta\psi)^{1/2}\right\rangle}{\left\langle 1/(\psi - \delta\psi)^{1/2}\right\rangle}$$

$$-\frac{\mu_0}{\sqrt{2B_y'}}\frac{\mathrm{d}\ln\eta}{\mathrm{d}x}j_z\int_{-\infty}^{\infty}\frac{\mathrm{d}\psi}{(\psi - \delta\psi)^{1/2}}\frac{\left\langle \xi_x/(\psi - \delta\psi)^{1/2}\right\rangle}{\eta\left\langle 1/(\psi - \delta\psi)^{1/2}\right\rangle}.$$

Multiplying by $\cos ky$ and averaging over y, this equation is reduced to

$$\frac{\left.\frac{\partial\delta\psi_1}{\partial x}\right|_{-\infty}^{\infty}}{\delta\psi_1}\delta\psi_1 = -\frac{2\mu_0}{\sqrt{2B_y'}}\left(\frac{1}{\eta}\frac{\partial\delta\psi_1}{\partial t} + \xi_1\frac{\mathrm{d}\ln\eta}{\mathrm{d}x}j_z\right)$$

$$\times\int_{-\infty}^{\infty}\mathrm{d}\psi\frac{\left\langle \cos ky/(\psi - \delta\psi)^{1/2}\right\rangle^2}{\left\langle 1/(\psi - \delta\psi)^{1/2}\right\rangle}.$$

Introducing the dimensionless parameter $\Delta' = \frac{\partial \delta\psi_1}{\partial x}\Big|_{-\infty}^{+\infty} / \delta\psi_1$, one obtains the tearing mode equation:

$$\Delta' = \frac{2\sqrt{2}\mu_0}{\eta}\frac{\partial x_T}{\partial t}A_0 + \frac{4\sqrt{2}\mu_0 j_z}{B_y'}\frac{\mathrm{d}\ln\eta}{\mathrm{d}x}\frac{\xi_1}{x_T}A_0, \tag{3.63}$$

where

$$A_0 = \int_{-1}^{+\infty} \mathrm{d}w \frac{\left\langle \cos ky/(w - \cos ky)^{1/2} \right\rangle^2}{\left\langle 1/(w - \cos ky)^{1/2} \right\rangle} = 0.55.$$

Equation (3.63) is the modified Rutherford equation.

Without the resistivity gradient effect, (3.63) indicates that the tearing modes are nonlinearly unstable as soon as $\Delta' > 0$. The last term on the right-hand side of (3.63) represents the current interchange effect [12]. When there is a current gradient in the system, the interchange-type modes can carry over the current and consequently a current sheet results in the singular layer. This term is inversely proportional to the island width x_T. Therefore, it is very significant to the birth (the small x_T case) of the magnetic islands. In the derivation, only the Ohmic current is taken into consideration. The current interchange effects can include other types of currents, such as the bootstrap current. In particular, we note that the bootstrap current gradient can be very steep in the pedestal region. Also, at the plasma edge the current flow patterns inside the last closed flux surface and in the scrape-off layer (SOL) are quite different. There are closed magnetic surfaces inside the last closed flux surface, while the field lines end with divertors in the SOL. Therefore, there is a current jump across the last closed flux surface. This causes the final term in (3.63) to become even more significant [13]. The final term can drive new types of tearing modes, referred to as the current interchange tearing modes. The current interchange tearing modes can help in understanding many experimental observations (see chapter 5), for example the conversion of resistive wall modes to tearing modes, etc.

3.4 Discussion: neoclassical tearing modes, etc

In this chapter, we have described the resistive MHD theory. There can be neoclassical or kinetic modifications to this theory. The neoclassical tearing mode (NTM) is the tearing mode driven by the missing bootstrap current. When a seed magnetic island is induced by other instabilities, the pressure profile inside the magnetic island is locally flattened and the pressure gradient is nearly absent. Consequently, the equilibrium bootstrap current is missing. The missing equilibrium current is equivalent to a nonlinear current perturbation, which can cause the island to grow [14, 15]. The NTM can be linearly stable but nonlinearly unstable. This implies that even if $\Delta' < 0$ and the current interchange effect is excluded from consideration, it can still develop. The current interchange mode discussed in section 3.3 is the natural seed for the NTMs, in particular noting that its drive is inversely proportional to the island width.

Next, let us discuss the kinetic effects on resistive MHD theory [7, 8]. The ballooning mode formalism in section 3.2 is employed for this discussion. Using the ordering estimates in (3.36), one can find from (3.33) that

$$D_1 \sim \frac{S}{\max\left\{\omega_{si}^2, \omega^2\right\}} \sim \frac{S}{\max\left\{v_{\|f}^2/R^2q^2, \omega^2\right\}}, \tag{3.64}$$

where S represents the right-hand side of (3.33), and $v_{\|f}$ is the parallel fluid velocity. Note that D_1 is inversely proportional to $v_{\|f}^2$. As discussed previously in this book, the particles originally located in a small cell cannot be expected to remain in the same cell after the time period of interest. Therefore, one can expect the parallel fluid velocity in (3.64) to be replaced by the individual ion velocity in the kinetic description, leading to

$$D_1 \sim \frac{S}{\max\left\{v_{\|i}^2/R^2q^2, \omega^2\right\}}. \tag{3.65}$$

In the low frequency regime $\omega_{si} \gg \omega$, equation (3.65) indicates that the ions with parallel velocities much smaller than the thermal velocity give rise to a much more significant contribution, of the order $v_{thi}^2/\max\{v_{\|i}^2, \omega^2\}$, than the majority of ions with parallel velocities of same order as the thermal velocity. Noting that the population of ions with parallel velocities much smaller than the thermal velocity is reduced only by a factor of $\max\{v_{\|i}, \omega\}/v_{thi}$, compared to the majority. Consequently, the contribution of the minority with small parallel velocities is of order $v_{thi}/\max\{v_{\|i}, \omega\}$ larger than that of the majority. This reflects the so-called small-parallel-ion-velocity enhancement in [16] and [17] in the ideal MHD case.

Taking into consideration this small-parallel-ion-velocity enhancement of D_1, one can see that the apparent mass term in (3.48)

$$M_t \sim \frac{v_{thi/Rq}}{\max\left\{v_{\|i}/Rq, \omega\right\}} M_c \tag{3.66}$$

becomes much larger. To rebuild the ordering balance of each term in (3.47), one has to scale down z in defining the outer region (i.e. enlarge ϵ_θ). Consequently, one has that $M_t \gg M_c$ in the definition of Q in the final term of (3.47) and $x \ll 1$ in the denominators in the various terms in (3.47). After this reduction the only remaining collisional effect becomes the collisional modification of the apparent mass term with the small-parallel-ion-velocity enhancement. This shows an important kinetic modification to the resistive MHD theory as pointed out in [7] and [8].

Bibliography

[1] Glasser A H, Greene J M and Johnson J L 1975 Resistive instabilities in general toroidal plasma configurations *Phys. Fluids* **18** 875–88

[2] Johnson J L and Greene J M 1967 Resistive interchanges and the negative v'' criterion *Plasma Phys.* **9** 611

[3] Newcomb W A 1960 Hydromagnetic stability of a diffuse linear pinch *Ann. Phys. NY* **10** 232–67
[4] Shi T, Wang H and Zheng L J 2013 private communication
[5] Chance M S *et al* 1979 MHD stability limits on high-beta tokamaks *Proc. 7th Int. Conf. Plasma Physics and Controlled Fusion Research (Innsbruck, Austria, 23–30 August, 1978)* vol 1 (Vienna: International Atomic Energy Agency) p 677
[6] Correa-Restrepo D 1982 Resistive ballooning modes in three-dimensional configurations *Z. Naturforsch.* A **37** 848
[7] Zheng L-J and Tessarotto M 1996 Collisional effect on the magnetohydrodynamic modes of low frequency *Phys. Plasmas* **3** 1029–37
[8] Zheng L-J and Tessarotto M 1995 Collisional ballooning mode dispersion relation in the banana regime *Phys. Plasmas* **2** 3071–80
[9] Connor J W, Hastie R J and Taylor J B 1979 High mode number stability of an axisymmetric toroidal plasma *Proc. R. Soc. A* **365** 1–17
[10] Abramowitz M and Stegun I A (ed) 1972 *Handbook of Mathematical Functions* (Washington, D.C.: National Bureau of Standards)
[11] Rutherford P H 1973 Nonlinear growth of the tearing mode *Phys. Fluids* **16** 1903–8
[12] Zheng L J and Furukawa M 2010 Current-interchange tearing modes: Conversion of interchange-type modes to tearing modes *Phys. Plasmas* **17** 052508
[13] Zheng L J and Furukawa M 2014 Peeling-off of the external kink modes at tokamak plasma edge *Phys. Plasmas* **21** 082515
[14] Carrera R, Hazeltine R D and Kotschenreuther M 1986 Island bootstrap current modification of the nonlinear dynamics of the tearing mode *Phys. Fluids* **29** 899–902
[15] Callen J D, Qu W X, Siebert K D, Carreras B A, Shaing K C and Spong D A 1979 Neoclassical MHD equations instabilities and transport in tokamaks *Proc. 11th Int. Conf. Plasma Physics and Controlled Nuclear Fusion Research (Kyoto, Japan, 13–20 November 1986)* vol 2 (Vienna: International Atomic Energy Agency) p 157
[16] Mikhailovsky A 1974 *Theory of Plasma Instabilities* (New York: Consultants Bureau)
[17] Zheng L-J and Tessarotto M 1994 Collisionless kinetic ballooning mode equation in the low-frequency regime *Phys. Plasmas* **1** 3928–35

Linjin Zheng

Chapter 4

Gyrokinetic theory

We have discussed the ideal and resistive MHD theories in previous chapters. In this chapter we describe the kinetic stability theory.

On one hand, we know that the particle motions along the magnetic field lines are not localized. Particles can be trapped by the equilibrium magnetic field, or resonate with waves, etc. In the MHD theories we have pointed out that the minority particles with small parallel velocities can sometimes play a dominant inertia role as compared with the thermal particles. All of these indicate the necessity of a kinetic description in the parallel direction.

On the other hand, we know that the particle localization perpendicular to the magnetic field lines is valid only in the lowest order. Due to the inhomogeneity of the magnetic field lines, charged particles can drift away from them. This is because the charged particles experience a different magnetic field strength in their gyromotions. This is the so-called finite Larmor radius (FLR) effect. One also requires a kinetic description to take into account the FLR effects.

To include kinetic effects the most efficient way is to invoke the gyrokinetic formalism. The Vlasov equation is a seven-dimensional (including time) partial differential equation. By introducing the particle motion (adiabatic) invariants and applying high gyrofrequency ordering, the gyrokinetic formalism effectively reduces the seven-dimensional substantial time derivative in the linearized Vlasov equation into a four-dimensional one and therefore greatly simplifies the calculation of kinetic effects. The classic electrostatic gyrokinetic formalism was developed in the 1960s [1, 2]. Later, the electrostatic gyrokinetics was extended to develop an electromagnetic gyrokinetics [3, 4]. The gyrokinetics theory was revisited recently in [5], with two major corrections: the inclusion of an equilibrium distribution function to a sufficient order and the retention of the gyrophase-dependent contributions of the perturbed distribution function. This newly formulated gyrokinetic theory recovers the ideal MHD and FLR effects which are missing in the previous formalism.

doi:10.1088/978-1-6270-5423-2ch4

In this chapter we first derive the electrostatic gyrokinetic theory and investigate the theory of ion temperature gradient modes. After that we derive the electromagnetic gyrokinetic equation and apply the equation to study the interchange and ballooning modes.

4.1 The general gyrokinetic formalism and equilibrium

In this section we describe the general gyrokinetic formalism, in particular the underlying mathematical considerations. The gyrokinetic equilibrium is also discussed. The initial equation is the Vlasov equation:

$$\frac{\mathrm{d}F}{\mathrm{d}t} = \left[\frac{\partial}{\partial t} + \mathbf{v}\cdot\nabla_x + \frac{e}{m_\rho}(\mathbf{E} + \mathbf{v}\times\mathbf{B})\cdot\nabla_v\right]F = 0, \tag{4.1}$$

where F is the distribution function, e is the charge, m_ρ represents the mass, and ∇_x and ∇_v denote the Laplace operators in the configuration $\mathbf{x}$ and velocity $\mathbf{v}$ spaces, respectively. We will also use the subscripts 'i' and 'e' to represent the corresponding quantities for ion and electron species, respectively.

In order to linearize the Vlasov equation (4.1), the distribution function F is decomposed to the equilibrium (F) and perturbed (δF) parts. The linearized Vlasov equations for equilibrium and perturbation become, respectively,

$$\bar{\mathcal{L}}_v F = 0, \tag{4.2}$$

$$\mathcal{L}_v \delta F = -\frac{e}{m_\rho}\delta\mathbf{a}\cdot\nabla_v F, \tag{4.3}$$

where

$$\bar{\mathcal{L}}_v = \mathbf{v}\cdot\nabla_x + \left(e/m_\rho\right)\mathbf{v}\times\mathbf{B}\cdot\nabla_v, \tag{4.4}$$

$$\mathcal{L}_v = -\mathrm{i}\omega + \mathbf{v}\cdot\nabla_x + \left(e/m_\rho\right)\mathbf{v}\times\mathbf{B}\cdot\nabla_v, \tag{4.5}$$

and $\delta\mathbf{a} = \delta\mathbf{E} + \mathbf{v}\times\delta\mathbf{B}$, with $\delta\mathbf{E}$ and $\delta\mathbf{B}$ the perturbed electric and magnetic fields. Alternatively, the scalar $\delta\phi$ and vector $\delta\mathbf{A}$ potentials can be used to represent the perturbed electromagnetic field: $\delta\mathbf{E} = -\nabla\delta\phi - \partial\delta\mathbf{A}/\partial t$ and $\delta\mathbf{B} = \nabla\times\delta\mathbf{A}$.

As in [4], we use the phase space coordinate transform method to derive the gyrokinetic equation. In the configuration space, one notes that the strong magnetic field causes charged particles to gyrate. This leads to the introduction of the so-called guiding center coordinate

$$\mathbf{X} = \mathbf{x} + \frac{1}{\Omega}\mathbf{v}\times\mathbf{e}_b,$$

where $\Omega = eB/m_\rho$ is the gyrofrequency. In the velocity space one can introduce the invariant particle energy $\varepsilon = v^2/2$, the adiabatic invariant magnetic moment μ and the gyrophase α as new coordinates. The particle velocity is decomposed as

$\mathbf{v} = \mathbf{e}_1(\mathbf{x})v_\perp \cos\alpha + \mathbf{e}_2(\mathbf{x})v_\perp \sin\alpha + \mathbf{e}_b(\mathbf{x})v_\parallel$, where $\mathbf{e}_b = \mathbf{B}/B$, and the unit vectors $\mathbf{e}_1$, $\mathbf{e}_2$ and $\mathbf{e}_b$ form an orthogonal system in which $\mathbf{e}_b = \mathbf{e}_1 \times \mathbf{e}_2$. In the new coordinate system $(\mathbf{X}, \mathbf{V}) \equiv (\mathbf{X}, \varepsilon, \mu, \alpha)$ the substantial time derivative $\mathrm{d}/\mathrm{d}t$ in the Vlasov equation (4.1) becomes

$$\frac{\mathrm{d}}{\mathrm{d}t} = \frac{\partial}{\partial t} + \dot{\mathbf{X}} \cdot \nabla_{\mathbf{X}} + \dot{\mu}\frac{\partial}{\partial \mu} + \dot{\alpha}\frac{\partial}{\partial \alpha}, \tag{4.6}$$

where the dot denotes $\mathrm{d}/\mathrm{d}t$. One can prove that

$$\begin{aligned}
\dot{\mathbf{X}} &= v_\parallel \mathbf{e}_b + \mathbf{v}_D, \\
\dot{\mu} &= \mathbf{v} \cdot \nabla_{\mathbf{x}}\mu + \Omega\mathbf{v} \times \mathbf{e_b} \cdot \nabla_{\mathbf{v}}\mu, \\
\dot{\alpha} &= -\Omega(\mathbf{X}) + \dot{\alpha}_1,
\end{aligned}$$

where

$$\begin{aligned}
\mathbf{v}_D &= \mathbf{v} \times \left(\mathbf{v} \cdot \nabla_x\right)\left(\mathbf{e}_b/\Omega\right), \\
\dot{\alpha}_1 &= \mathbf{v} \cdot \nabla_x\alpha + (1/\Omega)\mathbf{v} \times \mathbf{e}_b \cdot \nabla_x\Omega, \\
\nabla_x\alpha &= \left(\nabla_x\mathbf{e}_2\right) \cdot \mathbf{e}_1 + \left(v_\parallel/v_\perp^2\right)\nabla_x\mathbf{e}_b \cdot (\mathbf{v}_\perp \times \mathbf{e}_b).
\end{aligned} \tag{4.7}$$

In $\dot{\alpha}_1$, the final term results from the guiding center transform of the gyrofrequency $\Omega(\mathbf{x})$. This correction is required, since $\mathbf{v} \cdot \nabla_x\alpha \sim (1/\Omega)\mathbf{v} \times \mathbf{e}_b \cdot \nabla_x\Omega$. As pointed out in [5], this correction is necessary to recover the ideal MHD from the gyrokinetics.

Unlike for $\dot{\mathbf{X}}$ and $\dot{\alpha}$, one requires $\dot{\mu} = 0$ to a required order, so that the $\mathrm{d}\mu/\mathrm{d}t$ term in (4.6) can be eliminated. Since the magnetic moment is an adiabatic variable, one can solve this required equation order by order. Corrected to order ρ/L (Larmor radius/equilibrium scale length), the result is [4]

$$\mu = \mu_0 + \tilde{\mu}_1 + \bar{\mu}_1, \qquad \mu_0 = v_\perp^2/2B,$$

$$\tilde{\mu}_1 = -\mathbf{v}_\mathrm{d} \cdot \mathbf{v}_\perp/B - \left(v_\parallel/4\Omega B\right)\left(\mathbf{v}_\perp \times \mathbf{e}_b \cdot \nabla\mathbf{e}_b \cdot \mathbf{v}_\perp + \mathbf{v}_\perp \cdot \nabla\mathbf{e}_b \cdot \mathbf{v}_\perp \times \mathbf{e}_b\right),$$

$$\bar{\mu}_1 = -\left(v_\parallel\mu_0/\Omega\right)\mathbf{e}_b \cdot \nabla \times \mathbf{e}_b,$$

$$\mathbf{v}_\mathrm{d} = \mathbf{e}_b \times \left[\left(v_\perp^2/2\Omega\right)\nabla \ln B + \left(v_\parallel^2/\Omega\right)\boldsymbol{\kappa}\right].$$

Therefore, in the guiding center coordinates, the equilibrium and perturbed Vlasov equations (4.2) and (4.3) become, respectively,

$$\bar{\mathcal{L}}_g F_g = 0, \tag{4.8}$$

$$\mathcal{L}_g \delta F_g = -\frac{e}{m_\rho}\left[\delta\mathbf{a} \cdot \nabla_V + \frac{1}{\Omega}\left(\delta\mathbf{a} \times \mathbf{e}_b\right) \cdot \nabla_{\mathbf{X}}\right]F_g, \tag{4.9}$$

where

$$\bar{\mathcal{L}}_g = \dot{\mathbf{X}} \cdot \nabla_X + \dot{\alpha}\frac{\partial}{\partial \alpha},$$
$$\mathcal{L}_g = \frac{\partial}{\partial t} + \dot{\mathbf{X}} \cdot \nabla_X + \dot{\alpha}\frac{\partial}{\partial \alpha}. \tag{4.10}$$

We first solve the equilibrium gyrokinetic equation, (4.8). The solution for the linearized equation, (4.9), will be described in the following sections. Introducing L_p to represent the scale length of equilibrium pressure and ρ to represent the Larmor radius, we adopt the following ordering assumption: $\rho_i/L_p \ll 1$. We also introduce L_B to represent the scale length of the magnetic field line curvature. We do not impose an ordering between L_p and L_B. However, L_p and L_B are kept explicitly in the ordering analyses, in order to make the specific physical ordering explicit, should the large aspect ratio configuration or the transport barrier physics be dealt with.

Corrected to order $O(\rho/L_B)$, the equilibrium Vlasov equation, (4.8), becomes

$$\left(v_\parallel \mathbf{e}_b \cdot \nabla_X - \Omega\frac{\partial}{\partial \alpha}\right)F_{g0} = 0. \tag{4.11}$$

This gives the lowest order equilibrium distribution function

$$F_{g0} = F_{g0}\big(\mathbf{X}_\perp, \mu, \varepsilon\big). \tag{4.12}$$

For simplicity, we use the isotropic Maxwellian distribution function as the lowest order solution

$$F_{g0}(\Psi, \varepsilon) = n_0(\Psi)\left(\frac{m_\rho}{2\pi T(\Psi)}\right)^{3/2} \exp\left\{-\frac{m_\rho \varepsilon}{T(\Psi)}\right\}, \tag{4.13}$$

where Ψ is the guiding center correspondent of poloidal flux ψ, n_0 is the plasma density and T denotes the temperature.

Most of existing linear gyrokinetic theories before [5] used only the equilibrium solution of this order. As will be shown later, this treatment results in the loss of MHD effects, not to mention an excessive neglect of FLR effects. To derive the perturbed gyrokinetic equation ordering consistently, the next order equilibrium distribution function F_{g1} is required. The solution of the equilibrium gyrokinetic equation of next order has been described in the neoclassical transport theory [6]. After gyrophase averaging the next order gyrokinetic equation reads

$$v_\parallel \mathbf{e}_b \cdot \nabla_X F_{g1} + \mathbf{v}_\mathrm{d} \cdot \nabla_X F_{g0} = 0.$$

Noting that $\mathbf{v}_\mathrm{d} \cdot \nabla\Psi = v_\parallel \mathbf{e}_\mathbf{b} \cdot \nabla_X(v_\parallel m_\rho g/eB)$, one obtains

$$F_{g1} = -v_\parallel \frac{g}{\Omega}\frac{\partial F_{g0}}{\partial \Psi} + \mathrm{sign}(v_\parallel)\bar{F}_{g1}(\Psi, \mu, \varepsilon), \tag{4.14}$$

where $\bar{F}_{g1}$ is the integration constant [6]. One can easily obtain the ordering estimate for F_{g1},

$$\frac{F_{g1}}{F_{g0}} \sim \frac{\rho L_B}{L_p^2}. \tag{4.15}$$

Let us discuss briefly the necessity to retain the next order equilibrium solution for the MHD equilibrium. The necessity to include the equilibrium distribution function of the first order for perturbed gyrokinetics, as pointed out in [5], will be discussed in later sections. We note first that [6] shows that, with and only with the first order equilibrium distribution function in (4.14) included, can the MHD equilibrium equations (1.10) and (1.15) be recovered. Noting that F_{g1} does not contribute to the perpendicular current density due to being odd in $v_\parallel$, one has

$$\begin{aligned}
\mathbf{J}_\perp &= \sum_{\mathrm{i,e}} e \int \mathrm{d}^3 v \mathbf{v}_\perp F_{g0}(\Psi, \varepsilon) \\
&= \sum_{\mathrm{i,e}} e \int \mathrm{d}^3 v \mathbf{v}_\perp \frac{1}{\Omega} \mathbf{v} \times \mathbf{e}_b \cdot \nabla F_{g0}(\Psi, \varepsilon) \\
&= \frac{\mathbf{B} \times \nabla P}{B^2},
\end{aligned} \tag{4.16}$$

where transforming from the guiding center to particle coordinates has been performed in the second step. Equation (4.16) recovers the MHD equilibrium result in (1.10). Noting also that F_{g0} does not contribute to the parallel current density due to being even in $v_\parallel$, one has

$$\begin{aligned}
J_\parallel &= \sum_{\mathrm{i,e}} e \int \mathrm{d}^3 v v_\parallel F_{g1}(\Psi, \varepsilon) \\
&= \sum_{\mathrm{i,e}} e \int \mathrm{d}^3 v v_\parallel \left(-v_\parallel \frac{g}{\Omega} \frac{\partial F_{g0}}{\partial \Psi} + \mathrm{sign}(v_\parallel) \bar{F}_{g1}(\Psi, \mu, \varepsilon) \right) \\
&= -\frac{gP'}{B} - B \sum_{\mathrm{i,e}} e \int \mathrm{d}\mu \mathrm{d}\varepsilon \bar{F}_{g1}(\Psi, \mu, \varepsilon).
\end{aligned} \tag{4.17}$$

Complete determination of the parallel current needs to derive $\bar{F}_{g1}$, which has been achieved in the neoclassical transport theory [6]. However, as justified in [6], one can generally impose that the surface function $\sum_{\mathrm{i,e}} e \int \mathrm{d}\mu \mathrm{d}\varepsilon \bar{F}_{g1}(\Psi, \mu, \varepsilon) = g'$, by comparing (4.17) with (1.15). Equation (4.17) actually describes the so-called Pfirsch–Schlüter current, as well as the parallel Ohmic current. From the derivations of (4.16) and (4.17) one can see that the lowest order equilibrium distribution function in (4.12) does not produce the parallel current, although it gives rise to the diamagnetic current. The parallel equilibrium current can only be retained by including the first order solution in (4.14).

It is interesting to note that in the particle coordinates the total equilibrium distribution function ($F_{g0} + F_{g1}$) can be expressed in the form of the quasi-shifted Maxwellian $F_{g0}(\mathbf{x}_\perp, (\mathbf{v} - \mathbf{V}^k)^2/2)$. Here, $\mathbf{V}^k$ is a function of space and velocity, instead of the fluid velocity as in the Braginskii two-fluid theory [7]. Note that the Braginskii two-fluid equations, as well as the gyroviscous tensor, are obtained by assuming that the equilibrium distribution function is a shifted Maxwellian. Without retaining the first order equilibrium distribution function F_{g1}, the parallel component of $\mathbf{V}^k$ would vanish. These arguments show that, to recover the MHD, one has to keep the the first order equilibrium distribution function.

4.2 The electrostatic gyrokinetic equation

Now, we derive the electrostatic linear gyrokinetic equation from the general gyrokinetic formalism outlined in section 4.1. In the electrostatic case the gyrokinetic equation (4.9) reduces to

$$\left(v_\parallel \mathbf{e}_b \cdot \nabla_X - \mathrm{i}\omega + \mathbf{v}_D \cdot \nabla_X + \dot{\alpha}\frac{\partial}{\partial \alpha}\right)\delta F_g$$

$$= \frac{e}{m_\rho}\left[\nabla\delta\phi(\mathbf{x}) \cdot \nabla_V + \frac{1}{\Omega}\left(\nabla\delta\phi(\mathbf{x}) \times \mathbf{e}_b\right) \cdot \nabla_\mathbf{X}\right]F_g. \tag{4.18}$$

Note here that the first term on the right is generally larger than the second term, since the second term derives from the FLR modification. In some cases one has to consider the contribution of the next order equilibrium distribution function and the gyrophase-dependent part of the perturbed distribution function. Nevertheless, the reasons for keeping them are not as important as in the electromagnetic case in section 4.4. We thus skip over these issues and consider only the equilibrium distribution function of lowest order and the gyrophase-averaged gyrokinetic equation for the electrostatic case. This this consistent with the electrostatic drift waves to be described in the following section.

Extracting the adiabatic part of the perturbed distribution function,

$$\delta F_g = \frac{e}{m_\rho}\frac{\partial F_{g0}}{\partial \varepsilon}\delta\phi(\mathbf{x}) + \delta h(\mathbf{X}, \varepsilon, \mu, \alpha), \tag{4.19}$$

the gyrokinetic equation (4.18) is reduced to, after the gyrophase average,

$$\left(v_\parallel \mathbf{e}_b \cdot \nabla_X - \mathrm{i}\omega + \mathbf{v}_D \cdot \nabla_X - \Omega(\mathbf{X})\frac{\partial}{\partial \alpha}\right)\delta h(\mathbf{X})$$

$$= \frac{e}{m_\rho}\left(\mathrm{i}\omega\frac{\partial F_{g0}}{\partial \varepsilon} + \frac{1}{\Omega}\mathbf{e}_b \times \nabla_X F_{g0} \cdot \nabla_x\right)\delta\phi(\mathbf{x}). \tag{4.20}$$

Note that the most interesting case for electrostatic modes is that for the high-n modes. We therefore introduce the ballooning mode representation in section 2.4

and reduce further the electrostatic gyrokinetic equation. Note that in obtaining (4.20) we only make the guiding center coordinate transform for the distribution function and $\delta\phi(\mathbf{x})$ is still given in the $\mathbf{x}$ space. We introduce the following WKB ansatz and then transform it to the guiding center coordinate

$$\delta\phi(\mathbf{x}) = \delta\bar{\phi}(\mathbf{x}) \exp\{\mathrm{i}S(\mathbf{x}_\perp)\} = \delta\bar{\phi}(\mathbf{X}) \exp\{\mathrm{i}S(\mathbf{X}_\perp) - \mathrm{i}L\}, \tag{4.21}$$

where $L = \mathbf{k}_\perp \cdot \mathbf{v} \times \mathbf{b}/\Omega = k_\perp v_\perp \cos\alpha$, $\mathbf{k}_\perp = \nabla_X S$, and $\delta\bar{\phi}(\mathbf{x})$ contains the slow variations in $\mathbf{x}$. Here, it is only necessary to keep the first order correction from the guiding center transform for the fast variant part $S(\mathbf{x})$. Using the generating function of the Bessel function, one can prove that $\langle\exp\{-\mathrm{i}L\}\rangle = J_0(k_\perp\rho)$ ($\langle\cdots\rangle$ represents the gyrophase average). Equation (4.20) becomes, after gyrophase averaging,

$$\mathbf{v}_\parallel \cdot \nabla\delta h - \mathrm{i}(\omega - \omega_\mathrm{d})\delta h = -\frac{\mathrm{i}e}{T}F_{g0}\left(\omega - \omega_*^T\right)J_0\left(k_\perp\rho\right)\delta\phi(\mathbf{X}), \tag{4.22}$$

where the hat bars and gyrophase average angular brackets have been dropped to simplify the notation, and

$$\omega_\mathrm{d} = \mathbf{k}_\perp \cdot \mathbf{v}_\mathrm{d}$$

$$\omega_*^T = \omega_*\left[1 + \eta\left(\frac{m_\rho\varepsilon}{T} - \frac{3}{2}\right)\right], \qquad \eta = \frac{\mathrm{d}\ln T}{\mathrm{d}\ln n}, \qquad \omega_* = \frac{nT}{e}\frac{\mathrm{d}\ln n}{\mathrm{d}\psi}.$$

One can solve (4.22) and use the solution $\delta h(\mathbf{X}, \mathbf{V})$ to compute the density moment. The eigenvalue problem can then be determined by the quasi-neutrality condition:

$$\sum_{\mathrm{i,e}}\frac{n_0e^2}{T}\delta\phi(\mathbf{x}) - \sum_{\mathrm{i,e}}2\pi e\int \mathrm{d}E\,\mathrm{d}\mu\frac{B}{|v_\parallel|}(\delta h_+ + \delta h_-)J_0 = 0. \tag{4.23}$$

Here, the subscripts $\pm$ denote the sign of parallel velocity, 'i, e' in the summations refer to the ion and electron species, and J_0 in the second term results from the transformation from the guiding center space back to the configuration space.

The set of equations (4.22) and (4.23) can be used to study the electrostatic mode stability problem.

4.3 Electrostatic drift waves

With the electrostatic gyrokinetic equation derived in section 4.2, in this section we describe the electrostatic drift wave theory. In chapter 2 we discussed the ideal MHD mode spectrum. In particular, we noted that there exists the sound wave spectrum, which is governed by the equation of parallel motion (2.8). In fact, since ions and electrons have vastly different masses, their responses to the waves can be quite different. This difference leads to the excitation of the parallel electric field. Consequently, the parallel dynamics tends to have non-single-fluid features and is coupled to the so-called electrostatic drift waves. The coupling of electrostatic drift waves to the sound waves in the electromagnetic description has been addressed in [8]. The quasi-neutrality condition results in an additional fluid-type resonance in the electromagnetic mode dispersion relation. Therefore, the discussion of electrostatic drift waves is not only interesting for the low beta case, in which the electrostatic approximation applies, but also for the parallel waves along the actual (equilibrium and perturbed) magnetic field lines in the finite beta case.

We focus our investigation on the electrostatic ion temperature gradient (ITG) modes and extend the long wavelength theory in [9]. In particular, we describe the multiple region matching technique to obtain the analytical or semi-analytical dispersion relations and discuss the external modes, as well as the internal ones. Both the slab and toroidal branches are investigated. Like the interchange modes, the slab branch has its resonance on the mode rational surface ($m - nq = 0$). The toroidal branch is instead toroidal Alfvén eigenmode (TAE)-like and involves interaction between two nearby Fourier components, resonating at the middle of two neighboring rational surfaces. There is a counterpart for the slab branch in the slab model, but not for the toroidal branch. This is why they are called 'slab' and 'toroidal', although the theory presented here is a toroidal one.

We employ the ballooning mode representation for this investigation. The quasi-toroidal coordinates (r, θ, ϕ) as shown in figure 1.1 are used. A large aspect ratio is assumed. In this case, one has

$$\mathbf{e}_b \cdot \nabla = \frac{1}{Rq}\frac{\partial}{\partial \theta},$$

$$\omega_\mathrm{d} = \left(\frac{v_\perp^2}{2} + v_\parallel^2\right)(\cos\theta + s\theta\sin\theta), \qquad \bar{\omega}_\mathrm{d} = \frac{nq}{\Omega_c Rr}.$$

The electrostatic gyrokinetic equation (4.22) and the quasi-neutrality condition (4.23) are used as the set of initial equations. For the excitation of electrostatic waves, ions and electrons need to have different responses. This leads us to consider the comparable frequency regimes: $\omega \sim \omega_\mathrm{ti}$. In these frequency regimes ions resonate with waves, while electrons tend to behave adiabatically. We use ω_ti and ω_ti to denote the electron and ion transit frequencies, respectively.

4.3.1 The slab-like branch

In this subsection we describe the theory of the ITG modes of the slab-like branch. This is an extension to the theory in [9]. For this the following ordering scheme is assumed:

$$\omega \sim \omega_{*\mathrm{i}} \sim \epsilon_T^{1/2}\omega_{*\mathrm{i}}^T \sim \omega_\mathrm{ti}, \qquad k_\wedge\rho_\mathrm{i} \sim \epsilon_T^{1/2}, \qquad \epsilon_\theta \sim \epsilon_T^{1/4}, \tag{4.24}$$

where the parameter $\epsilon_T \ll 1$. Here, we see that the temperature gradient is ordered to be steeper than the density gradient. Like the resistive ballooning mode theory in section 2.4 two regions are considered: the inner region ($\theta \sim 1$) and outer region ($\theta \sim 1/\epsilon_\theta$). We solve the equations for the two regions separately and then match their solutions in the common area.

For the outer region, we use two-scale analysis (θ, $\epsilon_\theta z$) and introduce the following representation:

$$\delta\phi = \exp\{-\mathrm{i}\omega t\} \sum_{i=0}^{\infty} \epsilon_\theta^i \delta\phi^{(i)}\left(\theta; \epsilon_\theta z\right), \tag{4.25}$$

with $\phi^{(i)}$ are periodic functions of θ. The representation of the same type also holds for distribution functions.

Because the mode frequency is assumed to be much lower than the electron transit frequency, the electron response can be assumed to be adiabatic, i.e. $\delta h_\mathrm{e} = 0$. We can then focus on solving for the ion distribution function and drop the subscript 'i' for brevity. Also, because of the large aspect ratio assumption, we can neglect the trapped particle effects and assume that $v_\parallel$ is independent of θ. Introducing the representation in (4.25), the gyrokinetic equation (4.22) in the lowest order reads

$$\frac{v_\parallel}{qR}\frac{\partial}{\partial\theta}\delta h^{(0)} - \mathrm{i}\omega\delta h^{(0)} = -\mathrm{i}\frac{Ze}{T}F_m\left(\omega - \omega_{*\mathrm{i}}^T\right)\delta\phi^{(0)}, \tag{4.26}$$

where Z denotes the charge number. The solution of (4.26) contains only the slowly varying part:

$$\delta h^{(0)}(z) = \frac{Ze}{T}F_m H_0^S \delta\phi^{(0)}(z), \tag{4.27}$$

where $H_0^S = 1 - \omega_{*\mathrm{i}}^T/\omega$.

The gyrokinetic equation (4.22) in the first order is

$$\frac{v_\parallel}{qR}\frac{\partial}{\partial\theta}\delta h^{(1)} - \mathrm{i}\omega\delta h^{(1)} = -\mathrm{i}\frac{Ze}{T}F_m\left(\omega - \omega_{*\mathrm{i}}^T\right)\delta\phi^{(1)} - \frac{v_\parallel}{qR}\frac{Ze}{T}F_m\left(1 - \frac{\omega_{*\mathrm{i}}^T}{\omega}\right)\frac{\partial}{\partial z}\delta\phi^{(0)}$$

$$- \mathrm{i}\frac{Ze}{T}F_m\bar{\omega}_\mathrm{d}\left(1 - \frac{\omega_{*\mathrm{i}}^T}{\omega}\right)sz\sin\theta\delta\phi^{(0)}. \tag{4.28}$$

where (4.27) has been used to express $(v_\parallel/qR)\partial h^{(0)}/\partial z$.

According to the equation type, one can search for the solution of the following type

$$\{\delta h, \delta\phi\}^{(1)} = \{\delta h, \phi\}_v^{(1)}(\epsilon_\theta z) + \{\delta h, \delta\phi\}_c^{(1)}(\epsilon_\theta z)\cos\theta + \{\delta h, \delta\phi\}_s^{(1)}(\epsilon_\theta z)\sin\theta.$$

The solution of (4.28) can therefore be specified as follows

$$\begin{aligned}
\delta h_v^{(1)} &= \frac{Ze}{T}F_m\left(H_0^S\delta\phi_v^{(1)} + H_{z\parallel}^S\frac{\partial}{\partial z}\delta\phi^{(0)}\right),\\
\delta h_c^{(1)} &= \frac{Ze}{T}F_m H_a^S\delta\phi_c^{(1)},\\
\delta h_s^{(1)} &= \frac{Ze}{T}F_m\left(H_a^S\delta\phi_s^{(1)} + H_s^S sz\delta\phi^{(0)}\right),
\end{aligned} \tag{4.29}$$

where

$$H_{z\|}^S = \mathrm{i}\left(v_\|/qR\omega\right)\left(1 - \omega_{*\mathrm{i}}^T/\omega\right),$$

$$H_a^S = \omega\left(\omega - \omega_{*\mathrm{i}}^T\right)/\left[\omega^2 - \left(v_\|/qR\right)^2\right],$$

$$H_s^S = \bar{\omega}_\mathrm{d}\left(\omega - \omega_{*\mathrm{i}}^T\right)/\left[\omega^2 - \left(v_\|/qR\right)^2\right].$$

With the solution, (4.29), of the ion gyrokinetic equation at this order, the quasi-neutrality condition, (4.23), yields that

$$\delta\phi_v^{(1)} = \delta\phi_c^{(1)} = 0, \qquad \delta\phi_s^{(1)} = \Phi_s sz\delta\phi^{(0)}, \tag{4.30}$$

where

$$\Phi_s = \frac{\frac{4\pi}{n_0}\int \mathrm{d}E\mathrm{d}\mu \frac{B}{\left|v_\|\right|} F_m H_s^S}{1 + \frac{1}{Z^2\tau} - \frac{4\pi}{n_0}\int \mathrm{d}E\mathrm{d}\mu \frac{B}{\left|v_\|\right|} F_m H_a^S},$$

where $\tau = T_\mathrm{e}/T_\mathrm{i}$.

The gyrokinetic equation (4.22) in the second order reads

$$\begin{aligned}\frac{v_\|}{qR}\frac{\partial}{\partial\theta}\delta h^{(2)} - \mathrm{i}\omega\delta h^{(2)} &= -\mathrm{i}\frac{Ze}{T}F_m\left(\omega - \omega_{*\mathrm{i}}^T\right)\delta\phi^{(2)}\\ &+\mathrm{i}\frac{Ze}{T}F_m\left(\omega - \omega_{*\mathrm{i}}^T\right)\frac{n^2q^2}{4r^2}\frac{v_\perp^2}{\Omega_c^2}s^2z^2\delta\phi^{(0)}\\ &-\frac{v_\|}{qR}\frac{\partial}{\partial z}\delta h^{(1)} - \mathrm{i}\omega_D^g\delta h^{(1)}sz\sin\theta - \mathrm{i}\omega_D^n\delta h^{(0)}\cos\theta.\end{aligned} \tag{4.31}$$

Solving the gyrokinetic equation of this order, (4.31), inserting the solution into the quasi-neutrality condition, (4.23), and taking into consideration the remaining contribution of the zero order, one obtains the eigenvalue equation governing the ITG modes of the slab-like branch:

$$S_{zz}\frac{\mathrm{d}^2}{\mathrm{d}z^2}\delta\phi^{(0)} + \left(\frac{1}{Z^2\tau} + \frac{\omega_{*\mathrm{i}}}{\omega} + M_\| s^2z^2 + M_\perp s^2z^2\right)\delta\phi^{(0)} = 0, \tag{4.32}$$

where

$$S_{zz} = \frac{\mathrm{i}}{\omega}\frac{4\pi}{n_0}\int \mathrm{d}E\mathrm{d}\mu\frac{B}{\left|v_\|\right|}F_m\frac{v_\|}{qR}H_{z\|}^S = \frac{\left(v_T/qR\right)^2}{2\omega^2}\left(1 - \frac{\omega_{*\mathrm{p}}}{\omega}\right),$$

$$M_\| = \frac{1}{\omega}\frac{4\pi}{n_0}\int \mathrm{d}E\mathrm{d}\mu\frac{B}{\left|v_\|\right|}F_m\bar{\omega}_\mathrm{d}\left(H_a^s\Phi_s + H_s^S\right),$$

$$M_\perp = \frac{n^2q^2}{4r^2}\frac{v_T^2}{\Omega_c^2}\left(1 - \frac{\omega_{*\mathrm{p}}}{\omega}\right),$$

$$\omega_{*\mathrm{p}} = \omega_{*\mathrm{i}}\left(1 + \eta_\mathrm{i}\right).$$

Remarkably, here one also finds that the inertia effect results both from the perpendicular ($M_\perp$) and parallel ($M_\|$) motions, just as the apparent mass phenomenon for interchange modes described in section 2.3.

To construct the eigenvalue problem analytically, one needs to obtain the solution of (4.32). Letting $x = z/2a$, and

$$a = 2s\left(-\frac{M_\perp + M_\|}{S_{zz}}\right)^{1/2},$$

$$\nu = 2a\frac{1/Z^2\tau + \omega_{*\mathrm{i}}/\omega}{S_{zz}} - \frac{1}{2},$$

equation (4.32) is transformed to the Weber equation [10]:

$$\frac{\mathrm{d}^2\delta\phi}{\mathrm{d}x^2} + \left(\nu + \frac{1}{2} - \frac{x^2}{4}\right)\delta\phi = 0.$$

Furthermore, we introduce the transform $\delta\phi = x^{-1/2}w$ and $r = x^2/2$, this equation is transformed to the Whittaker equation [10]:

$$\frac{\mathrm{d}^2 w}{\mathrm{d}r^2} + \left(\frac{1/4 - \mu^2}{r^2} + \frac{\lambda}{r} - \frac{1}{4}\right)w = 0, \tag{4.33}$$

where $\mu = 1/4$, $\lambda = \nu/2 + 1/4$. The solution of (4.33) is

$$w = c_1 W_{\lambda,\mu}(r) + c_2 W_{-\lambda,\mu}(-r), \tag{4.34}$$

where $c_{1,2}$ are constants and $W_{\lambda,\mu}(r)$ is Whittaker's function, which is related to the Kummer function $M(a, b, r)$ as follows

$$W_{\lambda,\mu}(r) = \mathrm{e}^{-r/2}r^{1/2+\mu}M(1/2 + \mu - \lambda, 1 + 2\mu, r).$$

Using the asymptotic behavior of the Kummer function for the large argument, one can find that $W_{\lambda,\mu}(r)$ diverges as $r \to \infty$ whether $\Re e\{r\}$ is positive or not. Therefore, one has to set $c_1 = 0$ in (4.34).

Now, we can use the asymptotic behavior of the Kummer function for the small argument

$$M(a, b, r) = 1 + \frac{ar}{b} + \cdots$$

and the small argument expansion of the exponential function to obtain the small r asymptotic behavior for $W_{-\lambda,\mu}(-r)$. Transforming back to the original notation, one finds

$$\delta\phi \propto z^{1/2+2\mu}\left[1 + \left(\frac{1}{8a_\mathrm{s}b_\mathrm{s}} + \frac{1}{16a_\mathrm{s}^2}\right)z^2\right], \tag{4.35}$$

as $z \to 0$. Here, $a_\mathrm{s} = 1/2 + \mu + \lambda$ and $b_\mathrm{s} + 1 + 2\mu$.

The inner solution usually needs to be solved numerically. One can expect the outer limit of the inner solution to have the same type of behavior as (4.35):

$$\delta\phi \propto z^{1/2+2\mu}\left(1 + \Delta_s z^2\right), \tag{4.36}$$

where Δ_s specifies the ratio of small to large solutions.

Matching (4.35) with (4.36) one obtains the dispersion relation of the ITG modes of the slab branch

$$\Delta_s = \frac{1}{8a_s b_s} + \frac{1}{16a_s^2}. \tag{4.37}$$

As soon as the parameter Δ_s is determined, one can determine the eigenvalue from the dispersion relation (4.37). Alternatively, the condition in (4.35) can used as the boundary condition at ∞ for numerical shooting in the inner region to determine the eigenvalues.

In sections 2.3 and 2.5 we saw that the internal interchange mode has an external correspondence: the peeling mode. Likewise, we can also find the external electrostatic modes of the slab branch. Equation (4.37) is the dispersion relation for the internal modes, which is obtained by requiring $c_1 = 0$ in (4.34). For external modes, both constants c_1 and c_2 in (4.34) need to be kept in order to match the vacuum solution at the plasma–vacuum interface. The inverse transform from the ballooning mode representation space to configuration space is needed in order to find the matching condition at the plasma–vacuum interface. With the ratio of c_1 to c_2 determined by the vacuum solution, one can derive the dispersion relation of external modes through the remaining matching in the plasma region.

4.3.2 The toroidal branch

We describe the theory for the ITG modes of the toroidal-like branch in this subsection. This is an extension to the theory in [9]. For this the following optimal ordering scheme is adopted:

$$\omega \sim \omega_{*i}^{\cdots} \sim \omega_{ti}, \qquad k_\wedge \rho_i \sim \epsilon_a, \qquad \epsilon_\theta \sim \epsilon_a^{1/2}. \tag{4.38}$$

Here, it is noted that a much steeper temperature gradient compared to the density gradient is not necessarily required. Therefore, the modes of this branch do not need to be referred to as the ion temperature gradient modes. Unlike [9], we use the two-region matching method, the inner region ($\theta \sim 1$) and outer region ($\theta \sim 1/\epsilon_\theta$), to derive the dispersion relation.

We first investigate the outer region. Two-scale length (θ,$\epsilon_\theta z$) analyses are performed. The perturbed field and distribution functions are decomposed as follows

$$\delta\phi = \exp\{-\mathrm{i}\omega t\}\sum_{i=0}^{\infty}\epsilon_\theta^i\left(\delta\phi_c^{(i)}(\epsilon_\theta z)\cos\frac{\theta}{2} + \delta\phi_s^{(i)}(\epsilon_\theta z)\sin\frac{\theta}{2}\right). \tag{4.39}$$

Introducing the representation in (4.39), the gyrokinetic equation of lowest order, (4.22), is therefore reduced to

$$-\frac{v_\parallel}{2qR}h_c^{(0)} - \mathrm{i}\omega\delta h_s^{(0)} = -\mathrm{i}\frac{Ze}{T}F_m\left(\omega - \omega_{*\mathrm{i}}^T\right)\delta\phi_s^{(0)},$$

$$-\mathrm{i}\omega\delta h_c^{(0)} + \frac{v_\parallel}{2qR}\delta h_s^{(0)} = -\mathrm{i}\frac{Ze}{T}F_m\left(\omega - \omega_{*\mathrm{i}}^T\right)\delta\phi_c^{(0)}.$$

Theses two equations can be solved to yield

$$\begin{aligned}\delta h_c^{(0)} &= \frac{Ze}{T}F_m\left(H_\omega^T\,\delta\phi_c^{(0)} + H_\parallel^T\,\delta\phi_s^{(0)}\right),\\ \delta h_s^{(0)} &= \frac{Ze}{T}F_m\left(-H_\parallel^T\,\delta\phi_c^{(0)} + H_\omega^T\,\delta\phi_s^{(0)}\right),\end{aligned} \tag{4.40}$$

where

$$\begin{aligned}H_\omega^T &= \omega\left(\omega - \omega_{*\mathrm{i}}^T\right)H^T,\\ H_\parallel^T &= -\mathrm{i}\left(v_\parallel/2qR\right)\left(\omega - \omega_{*\mathrm{i}}^T\right)H^T,\\ H^T &= 1/\left[\omega^2 - \left(v_\parallel/2qR\right)^2\right].\end{aligned}$$

Inserting the solution of the ion gyrokinetic equation of this order, (4.40), into the quasi-neutrality condition, (4.23), one obtains, respectively:

$$D\left\{\delta\phi_c^{(0)}, \delta\phi_s^{(0)}\right\} = \epsilon_\theta, \tag{4.41}$$

where

$$D = 1 + \frac{1}{Z^2\tau} - \frac{4\pi}{n_0}\int \mathrm{d}E\mathrm{d}\mu\frac{B}{\left|v_\parallel\right|}\frac{F_m\omega\left(\omega - \omega_{*\mathrm{i}}^T\right)}{\omega^2 - \left(v_\parallel/2qR\right)^2}.$$

Note that in contrast to [9], we only require that $D \sim \epsilon_\theta$, instead of imposing $D=0$ as the dispersion relation.

The gyrokinetic equation of the first order in ϵ_θ reads

$$-\frac{v_\parallel}{2qR}\delta h_c^{(1)} - \mathrm{i}\omega\delta h_s^{(1)} = -\mathrm{i}\frac{Ze}{T}F_m\left(\omega - \omega_{*\mathrm{i}}^T\right)\delta\phi_s^{(1)} - \frac{v_\parallel}{qR}\frac{\partial}{\partial z}\delta h_s^{(0)} - \mathrm{i}\frac{1}{2}\bar{\omega}_\mathrm{d}\delta h_c^{(0)}sz,$$

$$-\mathrm{i}\omega\delta h_c^{(1)} + \frac{v_\parallel}{2qR}\delta h_s^{(1)} = -\mathrm{i}\frac{Ze}{T}F_m\left(\omega - \omega_{*\mathrm{i}}^T\right)\delta\phi_c^{(1)} - \frac{v_\parallel}{qR}\frac{\partial}{\partial z}\delta h_c^{(0)} - \mathrm{i}\frac{1}{2}\bar{\omega}_\mathrm{d}\delta h_s^{(0)}sz,$$

They can be solved to yield

$$\delta h_c^{(1)} = -H^T\left[\frac{1}{2}\left(\frac{v_\parallel}{qR}\right)^2\frac{\partial}{\partial z}\delta h_s^{(0)} + \mathrm{i}\omega\frac{v_\parallel}{qR}\frac{\partial}{\partial z}\delta h_c^{(0)} + \mathrm{i}\frac{1}{2}\bar{\omega}_{\mathrm{d}}\left(\frac{v_\parallel}{2qR}\delta h_c^{(0)} + \mathrm{i}\omega\delta h_s^{(0)}\right)sz\right],$$

$$\delta h_{0s}^{(1)} = H^T\left[\frac{1}{2}\left(\frac{v_\parallel}{qR}\right)^2\frac{\partial}{\partial z}\delta h_c^{(0)} - \mathrm{i}\omega\frac{v_\parallel}{qR}\frac{\partial}{\partial z}\delta h_s^{(0)} + \mathrm{i}\frac{1}{2}\bar{\omega}_{\mathrm{d}}\left(\frac{v_\parallel}{2qR}\delta h_s^{(0)} - \mathrm{i}\omega\delta h_c^{(0)}\right)sz\right].$$

$$(4.42)$$

Here, we have taken out the $\delta\phi_{0\,c,s}^{(1)}$ contribution, since their contributions are of the same type as (4.41).

Using (4.42), the quasi-neutrality condition of the first order yields

$$\begin{aligned} Z^T\frac{\mathrm{d}}{\mathrm{d}z}\delta\phi_s^{(0)} + \frac{1}{2}szS^T\delta\phi_s^{(0)} + D\delta\phi_c^{(0)} &= 0, \\ -Z^T\frac{\mathrm{d}}{\mathrm{d}z}\delta\phi_c^{(0)} + \frac{1}{2}szS^T\delta\phi_c^{(0)} + D\delta\phi_s^{(0)} &= 0, \end{aligned} \tag{4.43}$$

where the lowest order contribution in (4.41) has been included and

$$Z^T = -\omega\frac{4\pi}{n_0}\int \mathrm{d}E\mathrm{d}\mu\frac{B}{\left|v_\parallel\right|}\frac{v_\parallel^2}{q^2R^2}\left(\omega - \omega_{*\mathrm{i}}^T\right)H^{T2}F_m,$$

$$S^T = \frac{4\pi}{n_0}\int \mathrm{d}E\mathrm{d}\mu\frac{B}{\left|v_\parallel\right|}\bar{\omega}_{\mathrm{d}}\left(\omega - \omega_{*\mathrm{i}}^T\right)H^TF_m.$$

The set of partially differential equations, (4.43), can be combined to give the following equation:

$$\frac{\mathrm{d}^2\delta\phi_c^{(0)}}{\mathrm{d}z^2} + \left(A^T - B^{T^2}z^2\right)\delta\phi_c^{(0)} = 0, \tag{4.44}$$

where

$$A^T = \frac{2D^2 - sS^TZ^T}{2Z^{T^2}}, \qquad B^T = \frac{sS^T}{2Z^T}.$$

Similar to the slab branch treatment, we introduce $z = x/(2B^T)^{1/2}$ and $\nu = (A^T/B^T - 1)/2$, so that (4.44) can be transformed to the Weber equation [10]

$$\frac{\mathrm{d}^2\delta\phi}{\mathrm{d}x^2} + \left(\nu + \frac{1}{2} - \frac{x^2}{4}\right)\delta\phi = 0. \tag{4.45}$$

Furthermore, we introduce the transformation $\delta\phi = x^{-1/2}w$ and $r = x^2/2$, this equation is transformed to the Whittaker equation [10]:

$$\frac{d^2 w}{dr^2} + \left(\frac{1/4 - \mu^2}{r^2} + \frac{\lambda}{r} - \frac{1}{4}\right)w = 0, \tag{4.46}$$

where $\mu = 1/4$, $\lambda = \nu/2 + 1/4$. The solution of (4.46) is

$$w = c_1 W_{\lambda,\mu}(r) + c_2 W_{-\lambda,\mu}(-r), \tag{4.47}$$

where $c_{1,2}$ are constants and $W_{\lambda,\mu}(r)$ is Whittaker's function, which is related to the Kummer function $M(a, b, r)$ as follows

$$W_{\lambda,\mu}(r) = e^{-r/2} r^{1/2+\mu} M(1/2 + \mu - \lambda, 1 + 2\mu, r).$$

Using the asymptotic behavior of the Kummer function for the large argument, one can find that $W_{\lambda,\mu}(r)$ diverges as $r \to \infty$ whether $\Re e\{r\}$ is positive or not. Therefore, one has to set $c_1 = 0$ in (4.47).

Now, we can use the asymptotic behavior of the Kummer function for the small argument

$$M(a, b, r) = 1 + \frac{ar}{b} + \cdots$$

and the small argument expansion of the exponential function to obtain the small r asymptotic behavior for $W_{-\lambda,\mu}(-r)$. Transforming back to the original notations, one finds

$$\delta\phi \propto z^{1/2+2\mu}\left[1 + \left(\frac{1}{8a_t b_t} + \frac{1}{16a_t^2}\right)z^2\right], \tag{4.48}$$

as $z \to 0$. Here, $a_t = 1/2 + \mu + \lambda$ and $b_t = 1 + 2\mu$.

The inner solution usually needs to be solved numerically. One can expect the outer limit of the inner solution to have the same type of behavior as (4.48):

$$\delta\phi \propto z^{1/2+2\mu}\left(1 + \Delta_t z^2\right), \tag{4.49}$$

where Δ_t specifies the ratio of the small to large solutions.

Matching (4.48) with (4.49) one obtains the dispersion relation of the ITG modes of the toroidal branch

$$\Delta_t = \frac{1}{8a_t b_t} + \frac{1}{16a_t^2}. \tag{4.50}$$

As soon as the parameter Δ_t is determined, one can determine the eigenvalue from the dispersion relation (4.50). Alternatively, the condition in (4.48) can used as the boundary condition at ∞ for numerical shooting in the inner region to determine the eigenvalues.

Similarly to the slab branch case one can construct the external mode dispersion relation as well. For this both constants c_1 and c_2 in (4.47) need to be kept in order to match the vacuum solution at the plasma–vacuum interface. With the ratio of c_1 to c_2 determined by the vacuum solution, one can derive the dispersion relation of the external modes through the remaining matching in the plasma region.

4.4 The electromagnetic gyrokinetic equation

In this section we derive the linear electromagnetic gyrokinetic equation. In the conventional derivation only the lowest order equilibrium distribution function is kept and only the gyrophase-averaged part of the perturbed distribution function is solved. This leads to Pfisch–Schlüter current effects and a considerable part of the FLR effects are left out. Here, we explain the extended derivation in [5] to recover the Pfisch–Schlüter current effects and the missing FLR effects.

The general gyrokinetics formalism has been described in section 4.1. After the guiding center coordinate transform, the invariant ε and adiabatic invariant μ have been transformed out in (4.9). However, the gyrophase variable still remains in the equation. Unlike the conventional approach, in which only the gyrophase-averaged part is solved, we solve (4.9) using the Fourier decomposition method with respect to the gyrophase α. This method was previously used for the cyclotron resonance problem in [11]. Here, it is shown that solving for the gyrophase-dependent part is necessary even for the modes with frequencies lower than the cyclotron frequency.

To recover MHD, we adopt the MHD gauge $\delta\mathbf{A} = \boldsymbol{\xi} \times \mathbf{B}$, so that

$$\delta\mathbf{a} = -\nabla\delta\phi + \mathrm{i}\omega\boldsymbol{\xi} \times \mathbf{B}. \tag{4.51}$$

As discussed in [12], this representation can be used for the cases without field line reconnections. To extract the adiabatic parts of the perturbed distribution function and convective contribution, we define

$$\delta f = \frac{e}{m_\rho}\frac{\partial F_{g0}}{\partial\varepsilon}\delta\phi(\mathbf{x}) - \boldsymbol{\xi}(\mathbf{X}) \cdot \nabla F_{g0} + \delta G(\mathbf{X}, \varepsilon, \mu, \alpha). \tag{4.52}$$

Here, it is noted that the electrostatic potential is given in the configuration $\mathbf{x}$ space, while the field displacement $\boldsymbol{\xi}(\mathbf{X})$ is given in the guiding center space. These are to ensure the effectiveness of the extractions. Using (4.51) and (4.52), the linear gyrokinetic equation (4.9) is then reduced to

$$\mathcal{L}^1\delta G - \Omega(\mathbf{X})\frac{\partial\delta G}{\partial\alpha} = \mathcal{R}, \tag{4.53}$$

where $\mathcal{L}_g = -\Omega(\mathbf{X})(\partial/\partial\alpha) + \mathcal{L}^1$, and

$$\mathcal{L}^1 = -\mathrm{i}\omega + \dot{\mathbf{X}} \cdot \nabla_X + \dot{\alpha}_1(\partial/\partial\alpha),$$

$$\begin{aligned}
\mathcal{R} = &-\mathrm{i}\omega\frac{e}{m_\rho}\frac{\partial F_{g0}}{\partial\varepsilon}\mathbf{v}_\perp \cdot \delta\mathbf{A}(\mathbf{x}) + \mathbf{v}_D \cdot \nabla_X\left(\boldsymbol{\xi}(\mathbf{X}) \cdot \nabla F_{g0}\right) + \mathrm{i}\omega(\boldsymbol{\xi}(\mathbf{x}) - \boldsymbol{\xi}(\mathbf{X})) \cdot \nabla F_{g0} \\
&+ \mathrm{i}\left(\omega - \omega_{*\mathrm{i}}^T\right)\frac{e}{m_\rho}\frac{\partial F_{g0}}{\partial\varepsilon}\delta\phi(\mathbf{x}) + \frac{e}{m_\rho}\mathbf{v}_D \cdot \nabla_X\left(\frac{\partial F_{g0}}{\partial\varepsilon}\right)\delta\phi(\mathbf{x}) \\
&+ \frac{e}{m_\rho}\left(\frac{g}{B}\frac{\partial F_{g0}}{\partial\Psi} + \left|v_\parallel\right|\frac{1}{B}\frac{\partial\bar{F}_{g1}}{\partial\mu}\right)\mathbf{e}_b \times \mathbf{v}_\perp \cdot \delta\mathbf{B}(\mathbf{x}) + \frac{1}{B}\mathbf{e}_b \cdot \delta\mathbf{B}(\mathbf{x})\mathbf{v}_\perp \cdot \nabla_X F_{g0} \\
&+ v_\parallel\frac{e}{m_\rho}\frac{\partial F_{g1}}{\partial\varepsilon}\mathbf{e_b} \cdot \nabla_x\delta\phi(\mathbf{x}) + \frac{e}{m_\rho}\left(\frac{\partial F_{g1}}{\partial\varepsilon} + \frac{1}{B}\frac{\partial F_{g1}}{\partial\mu}\right)\mathbf{v}_\perp \cdot \left(\nabla_x\delta\phi(\mathbf{x}) - \mathrm{i}\omega\delta\mathbf{A}(\mathbf{x})\right) \\
&- v_\parallel\frac{1}{B}\delta\mathbf{B}(\mathbf{x}) \cdot \nabla_X F_{g0} + v_\parallel\mathbf{e}_b(\mathbf{x}) \cdot \nabla_X\left(\boldsymbol{\xi}(\mathbf{X}) \cdot \nabla F_{g0}\right).
\end{aligned} \tag{4.54}$$

Here, the inclusion of F_{g1} on the right-hand side of (4.9) has produced several new terms. Noting that the order of the first to second terms on the right-hand side of (4.9) is formally $L_\mathrm{p}/\rho_\mathrm{i}$ and the order of F_{g1} to F_{g0} is $\rho L_B/L_\mathrm{p}^2$ as given by (4.53), one can conclude that the F_{g1} contribution to the first term on the right-hand side of (4.9) needs to be kept in general for ordering consistency, as soon as the second term, the conventional reference term for FLR effects, on the right is kept.

To proceed further, the following ordering assumption is adopted

$$\mathcal{L}^1/\Omega \sim \epsilon, \tag{4.55}$$

as in the conventional gyrokinetic theory. Although the final term in (4.53) $-\Omega(\partial/\partial\alpha)$ is the dominant term in the gyrokinetic ordering in (4.55), this does not imply $\partial\delta G/\partial\alpha = 0$ to the lowest order, since one cannot presume the ordering between the terms on the left-hand side and the source terms on the right-hand side of (4.53). As will become apparent in the analyses described in the following derivation, the gyrophase-dependent part of the distribution function should be kept, as well as the gyrophase-independent part, for ordering consistency.

We use the Fourier decomposition method to solve (4.53),

$$\{\delta G, \mathcal{R}\} = \sum_k \left\{\delta G_k, \mathcal{R}_k\right\} \exp\{\mathrm{i}k\alpha\}.$$

With the ordering assumption in (4.55), equation (4.53) can be solved order by order. Expanding the perturbed distribution function as

$$\delta G_k = \delta G_k^{(0)} + \epsilon\delta G_k^{(1)} + \ldots, \tag{4.56}$$

one has

$$
\begin{aligned}
\delta G_k^{(0)} &= \mathrm{i}\frac{1}{k\Omega}\mathcal{R}_k, \quad \text{for } k \neq 0, \\
\delta G_0^{(0)} &= {\mathcal{L}_{00}^1}^{-1}\left(\mathcal{R}_0 - \sum_l{}' \mathcal{L}_{0l}^1 \delta G_l^{(0)}\right), \\
\delta G_k^{(1)} &= -\mathrm{i}\frac{1}{k\Omega}\sum_l \mathcal{L}_{kl}^1 \delta G_l^{(0)}, \quad \text{for } k \neq 0, \\
\delta G_0^{(1)} &= -{\mathcal{L}_{00}^1}^{-1}\sum_l{}' \mathcal{L}_{0l}^1 \delta G_l^{(1)}, \\
&\cdots,
\end{aligned}
\tag{4.57}
$$

where $\sum_l'$ represents a summation excluding $l=0$ and the matrix elements

$$
\mathcal{L}_{lk}^1 = \frac{1}{2\pi}\int_0^{2\pi} \mathrm{d}\alpha \exp\{-\mathrm{i}l\alpha\}\, \mathcal{L}^1 \exp\{\mathrm{i}k\alpha\}.
$$

In addition to the ion and electron distribution functions $\delta f_{\mathrm{i,e}}$, there are three independent field unknowns: two components of $\boldsymbol{\xi}$ and a scalar $\delta\phi$. Two components of the Ampere's law and the quasi-neutrality condition, together with the ion and electron gyrokinetic equations, are used to construct the complete set of equations. Ampere's law reads

$$
\delta \mathbf{J} \cdot \mathbf{e}_{1,2} = \sum_{\mathrm{i,e}} e \int \mathrm{d}^3 v \mathbf{v}_\perp \cdot \mathbf{e}_{1,2} \delta f(\mathbf{x}). \tag{4.58}
$$

The quasi-neutrality condition is

$$
\sum_{\mathrm{i,e}} e \int \mathrm{d}^3 v \delta f(\mathbf{x}) = 0. \tag{4.59}
$$

In (4.58) and (4.59), the velocity space integration is performed in the particle coordinates, instead of in the guiding center coordinates.

Here, we make some remarks about various approaches to constructing the basic set of equations. In the conventional approach (see for example [13]), one perpendicular component of Ampere's law and the vorticity equation are used to construct the basic set of equations. Instead, we use the two perpendicular components of Ampere's law directly for two reasons: first, from the MHD vorticity equation (2.9) one can see that both pressure and velocity moments need to be calculated to derive the kinetic vorticity equation. Note that calculating the velocity moment alone is equivalent to calculating the current density in Ampere's law. To avoid calculating two moments, a direct construction of the kinetic vorticity equation from the gyrokinetic equations has been used previously (see for example [13]), by applying the operator $\sum_{\mathrm{i,e}} e \int \mathrm{d}^3 v$ on the gyrophase-averaged gyrokinetic equation and using the quasi-neutrality condition for simplification. However, noting that the nabla operator in the term $\mathbf{v}_{\parallel} \cdot \nabla_{\mathbf{X}}$ in the gyrokinetic equation is in the guiding center coordinates, one cannot simply move $\mathbf{e}_b \cdot \nabla_{\mathbf{X}}$ out of the velocity space integration in a given particle coordinate without subtle elaboration for FLR modification. Note further that the term $\mathbf{v}_{\parallel} \cdot \nabla_{\mathbf{X}}$ produces the field line bending term of the shear Alfvén mode, which is generally much larger than the inertia term, in which the FLR effects appear. These show that, to obtain a vorticity equation in the particle coordinates, a backward transform for the term $\mathbf{v}_{\parallel} \cdot \nabla_{\mathbf{X}}$ is needed in order to retain the FLR effects consistently. This makes the vorticity equation approach non-trivial. This point has not been addressed previously in the derivation of the gyrokinetic vorticity equation. Second, we note that the vorticity equation alone is not sufficient for the completeness of the basic set of equations. At least one perpendicular component of Ampere's law is needed. Since the calculation of the two perpendicular components of Ampere's law is similar, our approach of using the two perpendicular components of Ampere's law directly is therefore a straightforward approach for constructing the basic set of equations.

Next, we solve the gyrokinetic equation and reduce Ampere's law (4.58) and the quasi-neutrality condition (4.59) to recover the linear ideal MHD equations. Although the theory outlined so far is applicable to arbitrary FLR ordering, we restrict ourselves here to only keeping the effects which are larger than or of the same order as the perpendicular inertia effect with diamagnetic frequency shift (i.e. ω^2 replaced by $\omega(\omega - \omega_{*\mathrm{i}}(1 + \eta_{\mathrm{i}})))$, with $\omega_{*\mathrm{i}} = -\mathrm{i}(T_{\mathrm{i}}/e_{\mathrm{i}}B)(\partial \ln n_0/\partial\psi)\mathbf{e}_b \times \nabla\psi \cdot \nabla$ and $\eta_{\mathrm{i}} = \partial \ln T_{\mathrm{i}}/\partial \ln n_0$. In the ordering analyses, we assume that $\omega \gtrsim \omega_{*\mathrm{i}}$. We also retain explicitly two different perpendicular wavelengths, using $\lambda_{\perp}$ and $\lambda_{\wedge}$ to denote, respectively, the perpendicular wavelength normal and tangential to the magnetic surface. Noting that $\omega_{*\mathrm{i}} \sim \rho_{\mathrm{i}}/\lambda_{\wedge}$ as compared to the shear Alfvén frequency, the assumption $\omega \sim \omega_{*\mathrm{i}}$ implies that effects of order $(\rho_{\mathrm{i}}/\lambda_{\wedge})^2$ are kept in the MHD equations. In addition, we assume that $\rho_{\mathrm{i}}/\lambda_{\perp} \ll 1$, $\lambda_{\wedge} \gtrsim \lambda_{\perp}$, $\lambda_{\wedge} \ll L_{\mathrm{p}}$, and $L_B \gtrsim L_{\mathrm{p}}$.

First, let us solve for the first harmonic solution of $\delta\tilde{G}$, using the perturbation method outlined in (4.56). The details are given in appendix A.1. The solution is obtained as follows

$$
\begin{aligned}
&\delta\tilde{G}_1(\mathbf{x}) \\
&= -\mathrm{i}\omega\frac{1}{B}\frac{\partial F_{g0}}{\partial\varepsilon}\mathbf{e}_b \times \mathbf{v}_{\perp} \cdot \delta\mathbf{A} + \frac{\omega^2}{\Omega}\frac{m_\rho}{T}F_{g0}v_{\perp}\big(-\sin\alpha\mathbf{e}_1 + \cos\alpha\mathbf{e}_2\big) \cdot \boldsymbol{\xi}(\mathbf{x}) \\
&\quad - \mathrm{i}\omega\frac{m_\rho}{T}F_{g0}v_{\perp}\cos\alpha\mathbf{e}_1 \cdot \boldsymbol{\xi} - \mathrm{i}\omega\frac{m_\rho}{T}F_{g0}\frac{3v_{\perp}^3}{8\Omega^2}B^{-3/2}\cos\alpha\big\{\mathbf{e}_1\mathbf{e}_1 + \mathbf{e}_2\mathbf{e}_2\big\} : \nabla\nabla\Big(B^{3/2}\mathbf{e}_1 \cdot \boldsymbol{\xi}\Big) \\
&\quad - \mathrm{i}\omega\frac{m_\rho}{T}F_{g0}\cos\alpha\Bigg\{\mathbf{e}_1 \cdot \nabla\mathbf{e}_1 \cdot \boldsymbol{\xi}\Bigg[-\frac{5v_{\perp}}{8\Omega^2}\mathbf{e}_1 \cdot \left(\frac{v_{\perp}^2}{2}\nabla\ln B + v_{\parallel}^2\boldsymbol{\kappa}\right) \\
&\quad + \frac{v_{\perp}^3}{8\Omega^2}\frac{\partial\ln n_0}{\partial\psi}\left[1 + \eta\left(\frac{m_\rho\varepsilon}{T} - \frac{5}{2}\right)\right] - \frac{v_{\perp}^3}{16\Omega^2}\mathbf{e}_1 \cdot (\nabla\ln B)\Bigg]
\end{aligned}
$$

$$+ \mathbf{e}_2 \cdot \nabla \mathbf{e}_1 \cdot \boldsymbol{\xi} \left[-\frac{5v_\perp}{8\Omega^2} \mathbf{e}_2 \cdot \left(\frac{v_\perp^2}{2} \nabla \ln B + v_\parallel^2 \boldsymbol{\kappa} \right) - \frac{9v_\perp^3}{16\Omega^2} \mathbf{e}_2 \cdot (\nabla \ln B) \right]$$

$$+ \frac{7v_\perp^3}{8\Omega^2} \left[(\mathbf{e}_1 \cdot \nabla \mathbf{e}_1) \cdot \nabla \mathbf{e}_1 \cdot \boldsymbol{\xi} + (\mathbf{e}_2 \cdot \nabla \mathbf{e}_2) \cdot \nabla \mathbf{e}_1 \cdot \boldsymbol{\xi} \right]$$

$$+ \mathbf{e}_1 \cdot \nabla \mathbf{e}_2 \cdot \boldsymbol{\xi} \left[\frac{v_\perp}{8\Omega^2} \mathbf{e}_2 \cdot \left(\frac{v_\perp^2}{2} \nabla \ln B + v_\parallel^2 \boldsymbol{\kappa} \right) + \frac{7v_\perp^3}{16\Omega^2} \mathbf{e}_2 \cdot (\nabla \ln B) \right]$$

$$+ \mathbf{e}_2 \cdot \nabla \mathbf{e}_2 \cdot \boldsymbol{\xi} \left[-\frac{v_\perp}{8\Omega^2} \mathbf{e}_1 \cdot \left(\frac{v_\perp^2}{2} \nabla \ln B + v_\parallel^2 \boldsymbol{\kappa} \right) - \frac{v_\perp^3}{8\Omega^2} \frac{\partial \ln n_0}{\partial \psi} \left[1 + \eta \left(\frac{m_\rho \varepsilon}{T} - \frac{5}{2} \right) \right] \right.$$

$$\left. + \frac{v_\perp^3}{16\Omega^2} \mathbf{e}_1 \cdot \left(\nabla \ln B \right) \right] + \frac{v_\perp^3}{8\Omega^2} \left[-(\mathbf{e}_2 \cdot \nabla \mathbf{e}_1) \cdot \nabla \mathbf{e}_2 \cdot \boldsymbol{\xi} + (\mathbf{e}_1 \cdot \nabla \mathbf{e}_2) \cdot \nabla \mathbf{e}_2 \cdot \boldsymbol{\xi} \right] \Bigg\}$$

$$- \mathrm{i}\omega \frac{m_\rho}{T} F_{g0} v_\perp \sin \alpha \mathbf{e}_2 \cdot \boldsymbol{\xi} - \mathrm{i}\omega \frac{m_\rho}{T} F_{g0} \frac{3v_\perp^3}{8\Omega^2} B^{-3/2} \sin \alpha \left\{ \mathbf{e}_1 \mathbf{e}_1 + \mathbf{e}_2 \mathbf{e}_2 \right\} : \nabla \nabla \left(B^{3/2} \mathbf{e}_2 \cdot \boldsymbol{\xi} \right)$$

$$- \mathrm{i}\omega \frac{m_\rho}{T} F_{g0} \sin \alpha \Bigg\{ \mathbf{e}_1 \cdot \nabla \mathbf{e}_1 \cdot \boldsymbol{\xi} \left[-\frac{v_\perp}{8\Omega^2} \mathbf{e}_2 \cdot \left(\frac{v_\perp^2}{2} \nabla \ln B + v_\parallel^2 \boldsymbol{\kappa} \right) + \frac{v_\perp^3}{16\Omega^2} \mathbf{e}_2 \cdot (\nabla \ln B) \right]$$

$$+ \mathbf{e}_2 \cdot \nabla \mathbf{e}_1 \cdot \boldsymbol{\xi} \left[\frac{v_\perp}{8\Omega^2} \mathbf{e}_1 \cdot \left(\frac{v_\perp^2}{2} \nabla \ln B + v_\parallel^2 \boldsymbol{\kappa} \right) - \frac{3v_\perp^3}{8\Omega^2} \frac{\partial \ln n_0}{\partial \psi} \left[1 + \eta \left(\frac{m_\rho \varepsilon}{T} - \frac{5}{2} \right) \right] \right.$$

$$\left. + \frac{7v_\perp^3}{16\Omega^2} \mathbf{e}_1 \cdot (\nabla \ln B) \right] + \frac{v_\perp^3}{8\Omega^2} \left[(\mathbf{e}_2 \cdot \nabla \mathbf{e}_1) \cdot \nabla \mathbf{e}_1 \cdot \boldsymbol{\xi} - (\mathbf{e}_1 \cdot \nabla \mathbf{e}_2) \cdot \nabla \mathbf{e}_1 \cdot \boldsymbol{\xi} \right]$$

$$+ \mathbf{e}_1 \cdot \nabla \mathbf{e}_2 \cdot \boldsymbol{\xi} \left[-\frac{5v_\perp}{8\Omega^2} \mathbf{e}_1 \cdot \left(\frac{v_\perp^2}{2} \nabla \ln B + v_\parallel^2 \boldsymbol{\kappa} \right) + \frac{5v_\perp^3}{8\Omega^2} \frac{\partial \ln n_0}{\partial \psi} \left[1 + \eta \left(\frac{m_\rho \varepsilon}{T} - \frac{5}{2} \right) \right] \right.$$

$$\left. - \frac{9v_\perp^3}{16\Omega^2} \mathbf{e}_1 \cdot (\nabla \ln B) \right]$$

$$+ \mathbf{e}_2 \cdot \nabla \mathbf{e}_2 \cdot \boldsymbol{\xi} \left[-\frac{5v_\perp}{8\Omega^2} \mathbf{e}_2 \cdot \left(\frac{v_\perp^2}{2} \nabla \ln B + v_\parallel^2 \boldsymbol{\kappa} \right) - \frac{v_\perp^3}{16\Omega^2} \mathbf{e}_2 \cdot (\nabla \ln B) \right]$$

$$+ \frac{7v_\perp^3}{8\Omega^2} \left[(\mathbf{e}_1 \cdot \nabla \mathbf{e}_1) \cdot \nabla \mathbf{e}_2 \cdot \boldsymbol{\xi} + (\mathbf{e}_2 \cdot \nabla \mathbf{e}_2) \cdot \nabla \mathbf{e}_2 \cdot \boldsymbol{\xi} \right] \Bigg\}$$

$$
\begin{aligned}
&- \mathrm{i}\omega\frac{|\nabla\psi|}{\Omega^2}\frac{\partial F_{g0}}{\partial\psi}\mathbf{v}_\perp\cdot\nabla(\mathbf{e}_1\cdot\boldsymbol{\xi}) - \mathrm{i}\frac{1}{B\Omega}\left(\omega-\omega_{*\mathrm{i}}^T\right)\frac{\partial F_{g0}}{\partial\varepsilon}\mathbf{v}_\perp\cdot\nabla\delta\phi(\mathbf{x})\\
&- \frac{1}{B}\mathbf{v}\times\mathbf{e}_b\cdot\nabla\left(\frac{\partial F_{g0}}{\partial\varepsilon}\right)\delta\phi(\mathbf{x}) - \frac{3}{8B\Omega^2}v_\perp^3\mathbf{e}_1\cdot\nabla\left(\frac{\partial F_{g0}}{\partial\varepsilon}\right)\sin\alpha(\mathbf{e}_1\mathbf{e}_1+\mathbf{e}_2\mathbf{e}_2):\nabla\nabla\delta\phi(\mathbf{x})\\
&- \left(\frac{g}{\Omega B}\frac{\partial F_{g0}}{\partial\psi} + \frac{1}{B^2}\frac{\partial\bar{F}_{g1}}{\partial\mu}\left|v_\parallel\right|\right)\mathbf{v}_\perp\cdot\delta\mathbf{B}(\mathbf{x})\\
&- \frac{3}{8\Omega^3}\left(\frac{g}{B}\frac{\partial F_{g0}}{\partial\psi} + \frac{e}{m_\rho}\left|v_\parallel\right|\frac{\partial\bar{F}_{g1}}{\partial\mu}\right)v_\perp^2(\mathbf{e}_1\mathbf{e}_1+\mathbf{e}_2\mathbf{e}_2):\nabla\nabla\left(\mathbf{e}_b\cdot\delta\mathbf{B}(\mathbf{x})\right)\\
&- \frac{|\nabla\psi|}{B\Omega}\frac{\partial F_{g0}}{\partial\psi}v_\perp\sin\alpha\mathbf{e}_b\cdot\delta\mathbf{B}(\mathbf{x})\\
&- \frac{3|\nabla\psi|}{8\Omega^3 B}\frac{\partial F_{g0}}{\partial\psi}v_\perp^3\sin\alpha(\mathbf{e}_1\mathbf{e}_1+\mathbf{e}_2\mathbf{e}_2):\nabla\nabla\left(\mathbf{e}_b\cdot\delta\mathbf{B}(\mathbf{x})\right)\\
&- \frac{v_\perp}{\Omega}\left(\mathbf{e}_1\sin\alpha-\mathbf{e}_2\cos\alpha\right)\cdot\nabla\boldsymbol{\xi}\cdot\nabla F_{g0}\\
&- \frac{v_\perp^3}{8\Omega^3}|\nabla\psi|\frac{\partial F_{g0}}{\partial\psi}\left(\mathbf{e}_1\mathbf{e}_1\mathbf{e}_1\sin\alpha+\mathbf{e}_1\mathbf{e}_2\mathbf{e}_2\sin\alpha-\mathbf{e}_1\mathbf{e}_1\mathbf{e}_2\cos\alpha-\mathbf{e}_2\mathbf{e}_2\mathbf{e}_2\cos\alpha\right)\\
&\quad\vdots\,\nabla\nabla\nabla(\boldsymbol{\xi}\cdot\mathbf{e}_1)\\
&- \frac{v_\perp^3}{8\Omega^3}\frac{\partial F_{g0}}{\partial\psi}\mathbf{e}_1\cdot\nabla(|\nabla\psi|)\left(3\mathbf{e}_1\mathbf{e}_1\sin\alpha+\mathbf{e}_2\mathbf{e}_2\sin\alpha-2\mathbf{e}_1\mathbf{e}_2\cos\alpha\right):\nabla\nabla(\boldsymbol{\xi}\cdot\mathbf{e}_1)\\
&- \frac{v_\perp^3}{8\Omega^3}\frac{\partial F_{g0}}{\partial\psi}\mathbf{e}_2\cdot\nabla(|\nabla\psi|)\left(2\mathbf{e}_1\mathbf{e}_2\sin\alpha-\mathbf{e}_1\mathbf{e}_1\cos\alpha-3\mathbf{e}_2\mathbf{e}_2\cos\alpha\right):\nabla\nabla(\boldsymbol{\xi}\cdot\mathbf{e}_1)\\
&- \frac{v_\perp^3}{8\Omega^3}|\nabla\psi|\,\mathbf{e}_1\cdot\nabla\left(\frac{\partial F_{g0}}{\partial\psi}\right)\left(3\mathbf{e}_1\mathbf{e}_1\sin\alpha+\mathbf{e}_2\mathbf{e}_2\sin\alpha-2\mathbf{e}_1\mathbf{e}_2\cos\alpha\right):\nabla\nabla(\boldsymbol{\xi}\cdot\mathbf{e}_1).
\end{aligned}
\tag{4.60}
$$

Next, we derive the gyrophase-averaged gyrokinetic equation. Summarizing the calculations in appendix A.2, one can obtain the gyrophase-averaged gyrokinetic equation

$$\left(\mathbf{v}_\parallel \cdot \nabla - \mathrm{i}\omega + \mathrm{i}\omega_\mathrm{d}\right)\delta G_0(\mathbf{X})$$
$$= -\mathrm{i}\omega\mu_0 B \frac{\partial F_{g0}}{\partial \varepsilon}\nabla_\perp \cdot \boldsymbol{\xi} - \mathrm{i}\omega \frac{\partial F_{g0}}{\partial \varepsilon}\left(\mu_0 B - v_\parallel^2\right)\boldsymbol{\kappa} \cdot \boldsymbol{\xi} + \mathrm{i}\omega_\mathrm{d}\boldsymbol{\xi} \cdot \nabla F_{g0}$$
$$+ \mathrm{i}\left(\omega - \omega_*^T\right)\frac{e}{m_\rho}\frac{\partial F_{g0}}{\partial \varepsilon}\delta\phi - \frac{v_\perp^2}{\Omega}\mathbf{e}_1 \cdot \nabla_X F_{g0}\mathbf{e}_2 \cdot \nabla\left(\frac{1}{B}\mathbf{e}_b \cdot \delta\mathbf{B}\right)$$
$$+ \frac{v_\parallel v_\perp^2}{\Omega^2}(\mathbf{e}_1\mathbf{e}_1 + \mathbf{e}_2\mathbf{e}_2) : \nabla\nabla\left(\mathbf{e}_b \cdot \nabla\boldsymbol{\xi} \cdot \nabla F_{g0}\right)$$
$$- \frac{v_\parallel v_\perp^2}{\Omega^2}\left[\left(\mathbf{e}_1 \cdot \nabla\mathbf{e}_b\right) \cdot \nabla\left(\mathbf{e}_1 \cdot \nabla\boldsymbol{\xi} \cdot \nabla F_{g0}\right) + \left(\mathbf{e}_2 \cdot \nabla\mathbf{e}_b\right) \cdot \nabla\left(\mathbf{e}_2 \cdot \nabla\boldsymbol{\xi} \cdot \nabla F_{g0}\right)\right], \tag{4.61}$$

where $\omega_\mathrm{d} = -\mathrm{i}\mathbf{v}_\mathrm{d} \cdot \nabla$.

Using the first harmonic solution $\delta\tilde{G}_1(\mathbf{x})$ of the gyrokinetic equation in (4.60) and the gyrophase-averaged distribution function $G_0(\mathbf{X})$ in (4.61), one can calculate two perpendicular components of the current density in Ampere's law in (4.58). The $\delta\tilde{G}_1(\mathbf{X})$ contribution has been given in appendix A.1. The $\delta G_0(\mathbf{x})$ contribution can be expressed as follows

$$\sum_{\mathrm{i,e}} e\int \mathrm{d}^3 v\mathbf{v}_\perp \cdot \mathbf{e}_{1,2}\delta G_0(\mathbf{X})\Big|_\mathbf{x} = \sum_{\mathrm{i,e}} e\int \mathrm{d}^3 v\mathbf{v}_\perp \cdot \mathbf{e}_{1,2}\frac{1}{\Omega}\mathbf{v} \times \mathbf{e}_b \cdot \nabla\delta G_0(\mathbf{x})$$
$$= -\sum_{\mathrm{i,e}} m_\rho \int \mathrm{d}^3 v\mu_0 \mathbf{e}_{2,1} \cdot \nabla\delta G_0(\mathbf{x}). \tag{4.62}$$

This is term that corresponds to the MHD plasma compressibility effect: $-\Gamma P\nabla \cdot \boldsymbol{\xi}$.

Collecting the contributions from the individual terms given in appendix A.1 and (4.62), one obtains the two components of Ampere's law in the kinetic description:

$$\mathbf{e}_1 \cdot \nabla \times \delta\mathbf{B}$$
$$= -\frac{gP'}{B^2}\mathbf{e}_1 \cdot \delta\mathbf{B} - g'\mathbf{e}_1 \cdot \delta\mathbf{B} + \frac{1}{B}\mathbf{e}_2 \cdot \nabla\left(P'|\nabla\psi|\,\mathbf{e}_1 \cdot \boldsymbol{\xi}\right)$$
$$- \sum_{\mathrm{i,e}} m_\rho \int \mathrm{d}^3 v\mu_0 \mathbf{e}_2 \cdot \nabla\delta G_0(\mathbf{x}) + \frac{\omega^2}{B}\rho_m \mathbf{e}_2 \cdot \boldsymbol{\xi} - \mathrm{i}\omega\frac{|\nabla\psi|}{B\Omega_\mathrm{i}}P_\mathrm{i}'\mathbf{e}_1 \cdot \nabla\left(\mathbf{e}_1 \cdot \boldsymbol{\xi}\right)$$
$$+ \mathrm{i}n_0 m_{\rho\mathrm{i}}\left[\omega - \omega_{*\mathrm{i}}\left(1 + \eta_\mathrm{i}\right)\right]\frac{1}{B^2}\mathbf{e}_1 \cdot \nabla\delta\phi$$
$$- \mathrm{i}\omega\frac{2n_0 Te_\mathrm{i}}{m_\rho}\frac{3}{4\Omega_\mathrm{i}^2}B^{-3/2}\left\{\mathbf{e}_1\mathbf{e}_1 + \mathbf{e}_2\mathbf{e}_2\right\} : \nabla\nabla\left(B^{3/2}\mathbf{e}_1 \cdot \boldsymbol{\xi}\right)$$

$$- \mathrm{i}\omega \frac{2n_0 Te_{\mathrm{i}}}{m_\rho} \left\{ \mathbf{e}_1 \cdot \nabla \mathbf{e}_1 \cdot \boldsymbol{\xi} \left[-\frac{1}{8\Omega_{\mathrm{i}}^2} \mathbf{e}_1 \cdot \left(6\nabla \ln B + \frac{5}{2}\boldsymbol{\kappa} \right) + \frac{|\nabla\psi|}{4\Omega_{\mathrm{i}}^2} \frac{\partial \ln n_0}{\partial \psi} \left(1 + \eta_{\mathrm{i}} \right) \right] \right.$$

$$+ \mathbf{e}_2 \cdot \nabla \mathbf{e}_1 \cdot \boldsymbol{\xi} \left[-\frac{1}{8\Omega_{\mathrm{i}}^2} \mathbf{e}_2 \cdot \left(14\nabla \ln B + \frac{5}{2}\boldsymbol{\kappa} \right) \right]$$

$$+ \frac{7}{4\Omega_{\mathrm{i}}^2} \left[\left(\mathbf{e}_1 \cdot \nabla \mathbf{e}_1 \right) \cdot \nabla \mathbf{e}_1 \cdot \boldsymbol{\xi} + \left(\mathbf{e}_2 \cdot \nabla \mathbf{e}_2 \right) \cdot \nabla \mathbf{e}_1 \cdot \boldsymbol{\xi} \right]$$

$$+ \mathbf{e}_1 \cdot \nabla \mathbf{e}_2 \cdot \boldsymbol{\xi} \frac{1}{8\Omega_{\mathrm{i}}^2} \mathbf{e}_2 \cdot \left(8\nabla \ln B + \frac{1}{2}\boldsymbol{\kappa} \right)$$

$$+ \mathbf{e}_2 \cdot \nabla \mathbf{e}_2 \cdot \boldsymbol{\xi} \left[-\frac{1}{16\Omega_{\mathrm{i}}^2} \mathbf{e}_1 \cdot \boldsymbol{\kappa} - \frac{|\nabla\psi|}{4\Omega_{\mathrm{i}}^2} \frac{\partial \ln n_0}{\partial \psi} \left(1 + \eta_{\mathrm{i}} \right) \right]$$

$$\left. + \frac{1}{4\Omega_{\mathrm{i}}^2} \left[-\left(\mathbf{e}_2 \cdot \nabla \mathbf{e}_1 \right) \cdot \nabla \mathbf{e}_2 \cdot \boldsymbol{\xi} + \left(\mathbf{e}_1 \cdot \nabla \mathbf{e}_2 \right) \cdot \nabla \mathbf{e}_2 \cdot \boldsymbol{\xi} \right] \right\}$$

$$- \frac{3}{2\Omega_{\mathrm{i}}} \left[\frac{g n_0 T_{\mathrm{i}}^2}{B^2 \Omega_{\mathrm{i}} m_{\rho\mathrm{i}}} \frac{\partial \ln n_0}{\partial \psi} \left(1 + 2\eta_{\mathrm{i}} \right) - 4\pi m_{\rho\mathrm{i}} \int \mathrm{d}\varepsilon \mathrm{d}\mu_0 \mu_0 \bar{F}_{g1} \right]$$

$$\times \left(\mathbf{e}_1\mathbf{e}_1 + \mathbf{e}_2\mathbf{e}_2 \right) : \nabla\nabla \left(\mathbf{e}_1 \cdot \delta \mathbf{B} \right)$$

$$+ \frac{n_0 e_{\mathrm{i}}}{2\Omega_{\mathrm{i}}^3} \left(\frac{T}{m_{\rho\mathrm{i}}} \right)^2 |\nabla\psi| \frac{\partial \ln n_0}{\partial \psi} (1 + 2\eta) (\mathbf{e}_1\mathbf{e}_1\mathbf{e}_2 + \mathbf{e}_2\mathbf{e}_2\mathbf{e}_2) \vdots \nabla\nabla\nabla \left(\boldsymbol{\xi} \cdot \mathbf{e}_1 \right)$$

$$+ \frac{n_0 e_{\mathrm{i}}}{\Omega_{\mathrm{i}}^3} \left(\frac{T}{m_{\rho\mathrm{i}}} \right)^2 \frac{\partial \ln n_0}{\partial \psi} (1 + 2\eta) \mathbf{e}_1 \cdot \nabla (|\nabla\psi|) \mathbf{e}_1\mathbf{e}_2 : \nabla\nabla \left(\boldsymbol{\xi} \cdot \mathbf{e}_1 \right)$$

$$+ \frac{n_0 e_{\mathrm{i}}}{2\Omega_{\mathrm{i}}^3} \left(\frac{T}{m_{\rho\mathrm{i}}} \right)^2 \frac{\partial \ln n_0}{\partial \psi} (1 + 2\eta) \mathbf{e}_2 \cdot \nabla (|\nabla\psi|) \left(\mathbf{e}_1\mathbf{e}_1 + 3\mathbf{e}_2\mathbf{e}_2 \right) : \nabla\nabla \left(\boldsymbol{\xi} \cdot \mathbf{e}_1 \right)$$

$$+ \frac{n_0 e_{\mathrm{i}}}{\Omega_{\mathrm{i}}^3} \left(\frac{T}{m_{\rho\mathrm{i}}} \right)^2 |\nabla\psi|^2 \left\{ \left(\frac{\partial \ln n_0}{\partial \psi} \right)^2 \left(1 + \frac{15}{2}\eta^2 + 4\eta \right) \right.$$

$$\left. + \frac{\partial^2 \ln n_0}{\partial \psi^2} (1 + 2\eta) + \frac{\partial \ln n_0}{\partial \psi} \left(2\frac{\partial \eta}{\partial \psi} - \eta \frac{7}{2} \frac{\partial \ln T}{\partial \psi} \right) \right\} \mathbf{e}_1\mathbf{e}_2 : \nabla\nabla \left(\boldsymbol{\xi} \cdot \mathbf{e}_1 \right), \qquad (4.63)$$

$$
\begin{aligned}
&\mathbf{e}_2 \cdot \nabla \times \delta\mathbf{B} \\
&= -\frac{gP'}{B^2}\mathbf{e}_2 \cdot \delta\mathbf{B} - g'\mathbf{e}_2 \cdot \delta\mathbf{B} - \frac{P'|\nabla\psi|}{B^2}\mathbf{e}_b \cdot \delta\mathbf{B} - \frac{1}{B}\mathbf{e}_1 \cdot \nabla\left(P'|\nabla\psi|\ \mathbf{e}_1 \cdot \boldsymbol{\xi}\right) \\
&\quad + \sum_{\mathrm{i,e}} m_\rho \int \mathrm{d}^3 v \mu_0 \mathbf{e}_1 \cdot \nabla\delta G_0(\mathbf{x}) - \frac{\omega^2}{B}\rho_m \mathbf{e}_1 \cdot \boldsymbol{\xi} - \mathrm{i}\omega\frac{|\nabla\psi|}{B\Omega_\mathrm{i}}P_\mathrm{i}'\mathbf{e}_2 \cdot \nabla\left(\mathbf{e}_1 \cdot \boldsymbol{\xi}\right) \\
&\quad + \mathrm{i}n_0 m_{\rho\mathrm{i}}\left[\omega - \omega_{*\mathrm{i}}\left(1 + \eta_\mathrm{i}\right)\right]\frac{1}{B^2}\mathbf{e}_2 \cdot \nabla\delta\phi \\
&\quad - \mathrm{i}\omega\frac{2n_0 Te_\mathrm{i}}{m_\rho}\frac{3}{4\Omega_\mathrm{i}^2}B^{-3/2}\left\{\mathbf{e}_1\mathbf{e}_1 + \mathbf{e}_2\mathbf{e}_2\right\} : \nabla\nabla\left(B^{3/2}\mathbf{e}_2 \cdot \boldsymbol{\xi}\right) \\
&\quad - \mathrm{i}\omega\frac{2n_0 T_\mathrm{i} e_\mathrm{i}}{m_{\rho\mathrm{i}}}\left\{-\mathbf{e}_1 \cdot \nabla\mathbf{e}_1 \cdot \boldsymbol{\xi}\frac{1}{16\Omega_\mathrm{i}^2}\mathbf{e}_2 \cdot \boldsymbol{\kappa}\right. \\
&\quad + \mathbf{e}_2 \cdot \nabla\mathbf{e}_1 \cdot \boldsymbol{\xi}\left[\frac{1}{8\Omega_\mathrm{i}^2}\mathbf{e}_1 \cdot \left(8\nabla \ln B + \frac{1}{2}\boldsymbol{\kappa}\right) - \frac{3|\nabla\psi|}{4\Omega_\mathrm{i}^2}\frac{\partial \ln n_0}{\partial\psi}\left(1 + \eta_\mathrm{i}\right)\right] \\
&\quad + \frac{1}{4\Omega_\mathrm{i}^2}\left[\left(\mathbf{e}_2 \cdot \nabla\mathbf{e}_1\right) \cdot \nabla\mathbf{e}_1 \cdot \boldsymbol{\xi} - \left(\mathbf{e}_1 \cdot \nabla\mathbf{e}_2\right) \cdot \nabla\mathbf{e}_1 \cdot \boldsymbol{\xi}\right] \\
&\quad + \mathbf{e}_1 \cdot \nabla\mathbf{e}_2 \cdot \boldsymbol{\xi}\left[-\frac{1}{8\Omega_\mathrm{i}^2}\mathbf{e}_1 \cdot \left(14\nabla \ln B + \frac{5}{2}\boldsymbol{\kappa}\right) + \frac{5|\nabla\psi|}{4\Omega_\mathrm{i}^2}\frac{\partial \ln n_0}{\partial\psi}\left(1 + \eta_\mathrm{i}\right)\right] \\
&\quad - \mathbf{e}_2 \cdot \nabla\mathbf{e}_2 \cdot \boldsymbol{\xi}\frac{1}{8\Omega_\mathrm{i}^2}\mathbf{e}_2 \cdot \left(6\nabla \ln B + \frac{5}{2}\boldsymbol{\kappa}\right) \\
&\quad \left. + \frac{7}{4\Omega_\mathrm{i}^2}\left[\left(\mathbf{e}_1 \cdot \nabla\mathbf{e}_1\right) \cdot \nabla\mathbf{e}_2 \cdot \boldsymbol{\xi} + \left(\mathbf{e}_2 \cdot \nabla\mathbf{e}_2\right) \cdot \nabla\mathbf{e}_2 \cdot \boldsymbol{\xi}\right]\right\} \\
&\quad - \frac{3}{2\Omega_\mathrm{i}}\left[\frac{gn_0 T_\mathrm{i}^2}{B^2\Omega_\mathrm{i} m_{\rho\mathrm{i}}}\frac{\partial \ln n_0}{\partial\psi}\left(1 + 2\eta_\mathrm{i}\right) - 4\pi m_{\rho\mathrm{i}} \int \mathrm{d}\varepsilon \mathrm{d}\mu_0 \mu_0 \bar{F}_{g1}\right] \\
&\quad \times \left(\mathbf{e}_1\mathbf{e}_1 + \mathbf{e}_2\mathbf{e}_2\right) : \nabla\nabla\left(\mathbf{e}_2 \cdot \delta\mathbf{B}\right) \\
&\quad - \frac{3|\nabla\psi|e_\mathrm{i} n_0}{2B\Omega_\mathrm{i}^3}\left(\frac{T_\mathrm{i}}{m_{\rho\mathrm{i}}}\right)^2\frac{\partial \ln n_0}{\partial\psi}\left(1 + 2\eta_\mathrm{i}\right)\left(\mathbf{e}_1\mathbf{e}_1 + \mathbf{e}_2\mathbf{e}_2\right) : \nabla\nabla\left(\mathbf{e}_b \cdot \delta\mathbf{B}\right) \\
&\quad - \frac{n_0 e_\mathrm{i}}{2\Omega_\mathrm{i}^3}\left(\frac{T}{m_{\rho\mathrm{i}}}\right)^2 |\nabla\psi|\ \frac{\partial \ln n_0}{\partial\psi}(1 + 2\eta)\left(\mathbf{e}_1\mathbf{e}_1\mathbf{e}_1 + \mathbf{e}_1\mathbf{e}_2\mathbf{e}_2\right) \vdots \nabla\nabla\nabla\left(\boldsymbol{\xi} \cdot \mathbf{e}_1\right)
\end{aligned}
$$

$$
\begin{aligned}
&-\frac{n_0 e_{\rm i}}{2\Omega_{\rm i}^3}\left(\frac{T}{m_{\rho \rm i}}\right)^2 \frac{\partial \ln n_0}{\partial \psi}(1+2\eta)\mathbf{e}_1 \cdot \nabla(|\nabla\psi|)\left(3\mathbf{e}_1\mathbf{e}_1+\mathbf{e}_2\mathbf{e}_2\right) : \nabla\nabla\left(\boldsymbol{\xi}\cdot\mathbf{e}_1\right) \\
&-\frac{n_0 e_{\rm i}}{\Omega_{\rm i}^3}\left(\frac{T}{m_{\rho \rm i}}\right)^2 \frac{\partial \ln n_0}{\partial \psi}(1+2\eta)\mathbf{e}_2 \cdot \nabla(|\nabla\psi|)\mathbf{e}_1\mathbf{e}_2 : \nabla\nabla\left(\boldsymbol{\xi}\cdot\mathbf{e}_1\right) \\
&-\frac{n_0 e_{\rm i}}{2\Omega_{\rm i}^3}\left(\frac{T}{m_{\rho \rm i}}\right)^2 |\nabla\psi|^2\left\{\left(\frac{\partial \ln n_0}{\partial \psi}\right)^2\left(1+\frac{15}{2}\eta^2+4\eta\right)\right. \\
&\left.+\frac{\partial^2 \ln n_0}{\partial \psi^2}(1+2\eta)+\frac{\partial \ln n_0}{\partial \psi}\left(2\frac{\partial \eta}{\partial \psi}-\eta\frac{7}{2}\frac{\partial \ln T}{\partial \psi}\right)\right\}(\mathbf{e}_1\mathbf{e}_1+\mathbf{e}_2\mathbf{e}_2) : \nabla\nabla\left(\boldsymbol{\xi}\cdot\mathbf{e}_1\right).
\end{aligned}
\tag{4.64}
$$

These two equations are the kinetic counterparts of the two perpendicular components, (2.6) and (2.7), of the MHD velocity momentum equations.

To study the MHD modes we can use $\nabla \cdot \delta\mathbf{j} = 0$ to construct the vorticity equation, with $\delta\mathbf{j} = \nabla \times \delta\mathbf{B}$. The calculations are tedious and are described in appendix A.2. The resulting kinetic vorticity equation is as follows:

$$
\begin{aligned}
&\nabla \cdot \frac{\mathbf{B}}{B^2} \times \rho_m \omega^2 \boldsymbol{\xi} \\
&= \mathbf{B}\cdot\nabla\frac{\mathbf{B}\cdot\delta\mathbf{J}}{B^2} + \delta\mathbf{B}\cdot\nabla\sigma - \mathbf{J}\cdot\nabla\frac{\mathbf{B}\cdot\delta\mathbf{B}}{B^2} - \nabla\times\frac{\mathbf{B}}{B^2}\cdot\nabla(\boldsymbol{\xi}\cdot\nabla P) \\
&+ \nabla\times\frac{\mathbf{B}}{B^2}\cdot\nabla\left(\sum_{\rm i,e}\frac{m_\rho}{2}\int {\rm d}^3 v v_\perp^2 \delta G_0(\mathbf{x})\right) - {\rm i}\omega\frac{|\nabla\psi|}{B\Omega_{\rm i}}P_{\rm i}'\nabla_\perp^2\left(\mathbf{e}_1\cdot\boldsymbol{\xi}\right) \\
&- {\rm i}\omega\frac{3}{2B\Omega_{\rm i}}|\nabla\psi|\, P_{\rm i}'\mathbf{e}_1\mathbf{e}_1 : \nabla\nabla\left(\mathbf{e}_1\cdot\boldsymbol{\xi}\right) \\
&+ {\rm i}\omega\frac{3n_0 T_{\rm i}}{B\Omega_{\rm i}}\mathbf{e}_2\cdot\left(\boldsymbol{\kappa}+\nabla\ln B-\mathbf{e}_1\cdot\nabla\mathbf{e}_1\right)\mathbf{e}_1\mathbf{e}_1 : \nabla\nabla\left(\mathbf{e}_2\cdot\boldsymbol{\xi}\right) \\
&- \frac{3\omega P_{\rm i}}{2B\Omega_{\rm i}}\left[\left(\mathbf{e}_2\cdot\nabla\mathbf{e}_1\right)\cdot\mathbf{e}_2\right]\mathbf{e}_1\cdot\nabla\left(\mathbf{e}_2\cdot\boldsymbol{\xi}\right) \\
&- {\rm i}\omega\frac{n_0 T_{\rm i}}{4B\Omega_{\rm i}}\mathbf{e}_2\cdot\left(8\nabla\ln B+\frac{1}{2}\boldsymbol{\kappa}\right)\mathbf{e}_1\mathbf{e}_1 : \nabla\nabla\left(\mathbf{e}_2\cdot\boldsymbol{\xi}\right) \\
&- {\rm i}\omega\frac{n_0 T_{\rm i}}{2B\Omega_{\rm i}}\mathbf{e}_1\left(\mathbf{e}_1\cdot\nabla\mathbf{e}_2\right) : \nabla\nabla\left(\mathbf{e}_2\cdot\boldsymbol{\xi}\right)
\end{aligned}
$$

$$+\frac{n_0 e_i}{2}\left(\frac{T}{m_{\rho i}}\right)^2 \frac{\partial \ln n_0}{\partial \psi}(1+2\eta_i)\left[\frac{|\nabla\psi|}{\Omega_i^3}\mathbf{e}_1\cdot(\mathbf{e}_1\cdot\nabla\mathbf{e}_2)-\nabla\cdot\left(\mathbf{e}_2\frac{|\nabla\psi|}{\Omega_i^3}\right)\right]$$

$$\times\,\mathbf{e}_1\mathbf{e}_1\mathbf{e}_1 \vdots \nabla\nabla\nabla(\boldsymbol{\xi}\cdot\mathbf{e}_1)$$

$$-\frac{n_0 e_i}{2\Omega_i^3}\left(\frac{T}{m_{\rho i}}\right)^2 \frac{\partial \ln n_0}{\partial \psi}(1+2\eta_i)(\mathbf{e}_1\cdot\nabla\,|\nabla\psi|)\mathbf{e}_1\mathbf{e}_1\mathbf{e}_2 \vdots \nabla\nabla\nabla(\boldsymbol{\xi}\cdot\mathbf{e}_1)$$

$$+\frac{n_0 e_i}{2\Omega_i^3}\left(\frac{T}{m_{\rho i}}\right)^2 |\nabla\psi|^2\left[\left(\frac{\partial \ln n_0}{\partial\psi}\right)^2\left(1+\frac{15}{2}\eta_i^{\,2}+4\eta_i\right)+\frac{\partial^2 \ln n_0}{\partial\psi^2}(1+2\eta_i)\right.$$

$$\left.+\frac{\partial \ln n_0}{\partial\psi}\left(2\frac{\partial\eta_i}{\partial\psi}-\eta_i\frac{7}{2}\frac{\partial\ln T}{\partial\psi}\right)\right]\mathbf{e}_1\mathbf{e}_1\mathbf{e}_2 \vdots \nabla\nabla\nabla(\boldsymbol{\xi}\cdot\mathbf{e}_1). \tag{4.65}$$

Here, one can see that the ideal MHD vorticity equation, (2.9), is recovered, except for the plasma compressibility effect, which is replaced by the kinetic correspondent, the fifth term on the right.

To determine $\delta\phi$, the quasi-neutrality condition (4.59) is needed. Inserting (4.52) into the quasi-neutrality condition (4.59) yields

$$\delta\phi(\mathbf{x}) = -\frac{1}{\sum_{i,e} n_0 e^2/T}\sum_{i,e} e\int d^3v\,\delta G_0. \tag{4.66}$$

Here, we have noted that for our ordering requirement no FLR expansion is needed for δG_0 and we have also considered the fact that the contribution from the gyrophase-dependent part of the perturbed distribution function is negligible. The parallel electric field effect enters into both the gyrophase-averaged gyrokinetic equation (4.61) and the two perpendicular components of Ampere's law, (4.63) and (4.64).

In the following, we discuss the gyrokinetic modifications to the ideal MHD. First, we discuss the $\delta\phi$ effect in the gyrokinetic equations (4.61). Inserting (4.66) into the terms containing $\delta\phi$ in (4.61), one can find that those terms become of the same order as the term $-i\omega\delta G_0$. This shows that the parallel electric field effect should be kept when the wave–particle resonance effect is taken into account. This fact is particularly relevant to the study of the kinetic stabilization of the resistive wall modes, in which the particle–wave resonance effect is critical. Comparing (4.63) and (4.64) with the MHD counterparts (2.6) and (2.7), one can see that δG_0 plays the role of $\nabla\cdot\boldsymbol{\xi}$. The sound wave resonance in the ideal MHD case in (2.8) is replaced by the wave–particle resonance in the kinetic description in (4.61). Note that the sound wave resonance is related to the sideband resonance [8]. Therefore, the parallel electric field effect comes mainly from the side band.

Next, we discuss the structural similarity of the basic set of equations between the ideal MHD and the current gyrokinetic descriptions. In the ideal MHD description, there are three unknowns: two components of $\boldsymbol{\xi}_\perp$ and the scalar $\nabla \cdot \boldsymbol{\xi}$. They are governed by the three projections of the MHD momentum equation: (2.6), (2.7) and (2.8). In the gyrokinetic description, the two perpendicular components of $\boldsymbol{\xi}_\perp$ remain, but the scalar $\nabla \cdot \boldsymbol{\xi}$ is replaced by the ion and electron gyrophase-averaged perturbed distribution functions δG_0. Note that in the kinetic description, $\boldsymbol{\xi}$ represents the field line displacement—a field variable; in the ideal MHD description, instead, $\boldsymbol{\xi}$ corresponds to the fluid velocity moment. The proportionality between $\nabla \cdot \boldsymbol{\xi}$ and δG_0 can be envisaged by the fluid continuity equation, noting that the convective part of the distribution function has been extracted in (4.52). Correspondingly, the two perpendicular projections of the MHD momentum equation, (2.6) and (2.7), are replaced by the two perpendicular projections of Ampere's law: (4.63) and (4.64); the parallel projection of the MHD momentum equation (2.8) is replaced by the gyrophase-averaged gyrokinetic equations for the ion and electron species (4.61). In the kinetic description, there is one more unknown $\delta\phi$, describing the parallel electric field effect. $\delta\phi$ is governed by the quasi-neutrality condition in (4.66).

Furthermore, let us discuss term by term the correspondences between the ideal MHD and the current gyrokinetic descriptions. Comparing the two perpendicular components of the ideal MHD momentum equations (2.6) and (2.7) with the two projections of Ampere's law in the kinetic description (4.63) and (4.64), one can see that all the MHD terms in (2.6) and (2.7), except for the plasma compressibility term (proportional to $\nabla \cdot \boldsymbol{\xi}$), are recovered in the kinetic equations (4.63) and (4.64). Note that the MHD fluid description is based on the particle spatial localization assumption. In the collisionless Vlasov equation description, however, the spatial localization can only be expected in the perpendicular direction due to the strong magnetic field, while particles can move freely in the parallel direction. Therefore, a fully kinetic description is needed in the parallel direction. Consequently, the terms due to the plasma compressibility effect in the ideal MHD description (the fourth term of (2.6) and the fifth term of (2.7)) are replaced by the kinetic moments for plasma compressibility (the fourth term of (4.63) and the fifth term of (4.64), respectively). Interestingly, there is also a structural similarity between the MHD equation of the parallel motion, (2.8), and the ion gyrophase-averaged gyrokinetic equation (4.61). Note that the kinetic plasma compressibility terms in (4.63) and (4.64) depend only on the even part (with respect to the parallel velocity) of the gyrophase-averaged distribution function. The gyrophase-averaged gyrokinetic equation governing the even part can be derived from (4.61) as given in (A.11) in appendix (A.2). To recover the MHD, the limit $\omega \gg \omega_{\mathrm{d}}, \omega_*$ should be taken. In this limit, ω_{d} on the left-hand side of (4.61) (or (A.11)) can be dropped and only the first two terms on the right-hand side of (4.61) need to be kept. Noting that $m_{\rho \mathrm{i}} \gg m_{\rho \mathrm{e}}$, only the ion distribution function needs to be kept. With these simplifications, the term by term correspondences between the MHD equation of parallel motion, (2.8) and the ion gyrokinetic equation (4.61) (with (A.11) used to construct the even part of the distribution function) are obvious, with the particle velocity replaced by the proper thermal velocity, noting the fact that the second term on the right-hand side of (4.61) vanishes with the thermal velocity replacement. It should be pointed out that the recovery of the similarity between the MHD and kinetic parallel descriptions is realized up to the order of the second term on the right-hand side of (4.61). This is particularly relevant for low frequency MHD modes, for which one usually has $\nabla \cdot \boldsymbol{\xi}_{\perp} \sim -2\boldsymbol{\kappa} \cdot \boldsymbol{\xi}$ and therefore the first and second terms on the right-hand side of (4.61) become of the same order.

In the collisionless kinetic description, the ions and electrons move individually along the field lines, instead of collectively as a fluid element. The different responses of the ions and electrons to the electromagnetic perturbations cause charge separation and thus the excitation of the parallel electric field. This causes the electrostatic scalar potential $\delta\phi$ to appear in the description of parallel motion through (4.61).

In addition, in the MHD description, the perpendicular motion of a fluid element is regarded to be the same as the field line displacement. In the kinetic description, however, the perpendicular motion of a fluid element is considered to be different from the field line displacement due to the FLR effects. The FLR effects lead to the changes: $\omega^2 \rightarrow \omega[\omega - \omega_{*i}(1 + \eta_i)]$ in the inertia terms (the fifth and sixth terms on the right-hand side of (4.63) and the sixth and seventh terms on the right-hand side of (4.64)) and also contributes additional FLR effect terms in (4.63), (4.64) and (4.61). Compare this to the existing gyrokinetic theories, in which the only FLR effect in this order is $\omega^2 \rightarrow \omega[\omega - \omega_{*i}(1 + \eta_i)]$ in the inertia terms [13]. In the current case, however, one can see that there are several additional FLR effect related terms in (4.63), (4.64) and (4.61).

The first two MHD terms on the left-hand sides of both (4.63) and (4.64) also are missing in the conventional gyrokinetic formulation. They are recovered here by taking into account the first order correction of the equilibrium distribution function and the gyrophase-dependent part of the perturbed distribution function. These two MHD terms result from the $\mathbf{J}_0 \times \delta\mathbf{B}$ term in the perpendicular force balance equations (2.6) and (2.7). Using only the Maxwellian equilibrium distribution function in the conventional gyrokinetic formulation, one cannot retain both the parallel Ohmic and Pfirsch–Schlüter currents.

4.5 FLR effects on the interchange modes

In this section we derive the kinetic singular layer equation, using the gyrokinetic formalism developed in section 4.4. Note that in the kinetic description, the field line displacement $\boldsymbol{\xi}$ (or the vector potential) is fluid-like, but one needs to include the FLR effects in the perpendicular direction. However, the parallel dynamics cannot be described by the parallel fluid velocity and the gyrokinetic description becomes necessary.

The derivation of the kinetic singular layer equation is similar to the ideal MHD equation in section 2.3. Only one thing needs to be noted. In section 2.3 we have assumed a low frequency regime, so that $\langle \delta\hat{P} \rangle \sim \Gamma P \langle \nabla \cdot \boldsymbol{\xi} \rangle$ is of order ϵ^2. Since we assume that $\omega \sim \omega_*$ in the current kinetic description, one can no longer use the low frequency assumption to rule out the $\langle \delta \hat{P} \rangle$ contribution. Consequently, in the current kinetic description we assume $\langle \delta \hat{P} \rangle$ to be of order ϵ, instead of ϵ^2. Noting that there is no parallel component for $\boldsymbol{\xi}$, one can use the following decompositions:

$$\boldsymbol{\xi} = \xi \frac{\nabla V}{|\nabla V|^2} + \mu \frac{\mathbf{B} \times \nabla V}{B^2},$$

$$\delta \mathbf{B} = b \frac{\nabla V}{|\nabla V|^2} + v \frac{\mathbf{B} \times \nabla V}{B^2} + \tau \frac{\mathbf{B}}{B^2}.$$

The ordering assumptions are as follows:

$$x \sim \epsilon, \qquad \frac{\partial}{\partial V} \sim \epsilon^{-1}, \qquad \frac{\partial}{\partial u} \sim \frac{\partial}{\partial \theta} \sim 1,$$

$$\xi = \epsilon \xi^{(1)} + \cdots, \qquad \mu = \mu^{(0)} + \cdots,$$

$$\langle \delta \hat{P} \rangle = \epsilon \langle \delta \hat{P} \rangle^{(1)} + \cdots, \qquad \widetilde{\delta \hat{P}} = \epsilon^2 \widetilde{\delta \hat{P}}^{(2)} + \cdots,$$

$$b = \epsilon^2 b^{(2)} + \cdots, \qquad v = \epsilon v^{(1)} + \cdots, \qquad \tau = \epsilon \tau^{(1)} + \cdots,$$

where $\epsilon \ll 1$ is the small parameter and $\delta \hat{P}$ is the non-convective part of perturbed plasma pressure

$$\delta \hat{P} = \sum_{i,e} \frac{m_\rho}{2} \int d^3 v \, v_\perp^2 \delta G_0(\mathbf{x}). \quad (4.67)$$

These ordering assumptions are similar to those in [14], except we use $\delta \hat{P}$ as the unknown to replace the parallel fluid displacement v. Here, we note that it can be proved *a posteriori* that $\langle \delta \hat{P} \rangle$ is one order larger than $\widetilde{\delta \hat{P}}$.

Since the derivation resembles the ideal MHD case in section 2.3, we only outline the key steps here. The condition that $\delta \mathbf{B}$ is divergence-free yields

$$\frac{\partial b^{(2)}}{\partial x} + \frac{1}{\Xi} \frac{\partial}{\partial u} v^{(1)} + \frac{J'}{P'} \frac{\partial v^{(1)}}{\partial \theta} - \frac{\chi'}{P'} \frac{\partial \sigma v^{(1)}}{\partial \theta} + \chi' \frac{\partial}{\partial \theta} \frac{\tau^{(1)}}{B^2} = 0. \quad (4.68)$$

After the surface average it gives

$$\frac{\partial \bar{b}^{(2)}}{\partial x} + \frac{1}{\Xi} \frac{\partial \bar{v}^{(1)}}{\partial u} = 0. \quad (4.69)$$

The two significant orders of the field representation $\delta \mathbf{B} = \nabla \times \boldsymbol{\xi} \times \mathbf{B}$ in the ∇V-direction are

$$0 = \chi' \frac{\partial \xi^{(1)}}{\partial \theta}, \tag{4.70}$$

$$b^{(2)} = \chi' \frac{\partial \xi^{(2)}}{\partial \theta} + \frac{\Lambda x}{\Xi} \frac{\partial \xi^{(1)}}{\partial u}, \tag{4.71}$$

From (4.71) and (4.68) one obtains

$$-\chi' \frac{\partial^2 \xi^{(2)}}{\partial \theta \partial x} = \frac{1}{\Xi} \frac{\partial \tilde{\nu}^{(1)}}{\partial u} + \frac{J'}{P'} \frac{\partial \nu^{(1)}}{\partial \theta} - \frac{\chi'}{P'} \frac{\partial \sigma \nu^{(1)}}{\partial \theta} + \chi' \frac{\partial}{\partial \theta} \frac{\tau^{(1)}}{B^2}. \tag{4.72}$$

The ∇u component of $\delta \mathbf{B} = \nabla \times \boldsymbol{\xi} \times \mathbf{B}$, in the lowest order, yields

$$\chi' \frac{\partial \mu^{(0)}}{\partial \theta} = 0. \tag{4.73}$$

To satisfy the parallel component of $\delta \mathbf{B} = \nabla \times \boldsymbol{\xi} \times \mathbf{B}$ along the magnetic field line, one must require that

$$\left(\nabla \cdot \boldsymbol{\xi}_\perp\right)^{(0)} + 2\boldsymbol{\kappa} \cdot \boldsymbol{\xi}^{(0)} = \frac{\partial \xi^{(1)}}{\partial x} + \frac{1}{\Xi} \frac{\partial \mu^{(0)}}{\partial u} = 0. \tag{4.74}$$

Taking into consideration the $\langle \delta \hat{P} \rangle$ contribution, the leading order of (4.63) and (4.64), the correspondents to the perpendicular MHD momentum equations, becomes

$$\tau^{(1)} + \delta P^{(1)} = \tau^{(1)} - P' \xi^{(1)} + \left\langle \delta \hat{P} \right\rangle^{(1)} = 0. \tag{4.75}$$

The first order of the kinetic vorticity equation, (4.65), becomes

$$\chi' \frac{\partial}{\partial \theta} \left(\frac{|\nabla V|^2}{B^2} \frac{\partial \nu^{(1)}}{\partial x} \right) + \chi' \frac{\partial \sigma}{\partial \theta} \frac{\partial \xi^{(1)}}{\partial x} - \frac{\chi'}{P'} \frac{\partial \sigma}{\partial \theta} \frac{\partial \left\langle \delta \hat{P}^{(1)} \right\rangle}{\partial x} = 0.$$

The solution of this equation is

$$\begin{aligned} \nu^{(1)} &= -\left(\frac{B^2 \sigma}{|\nabla V|^2} - \frac{\left\langle B^2 \sigma / |\nabla V|^2 \right\rangle}{\left\langle B^2 / |\nabla V|^2 \right\rangle} \frac{B^2}{|\nabla V|^2} \right) \xi^{(1)} - \Lambda \frac{B^2 / |\nabla V|^2}{\left\langle B^2 / |\nabla V|^2 \right\rangle} \frac{\partial}{\partial x} \left(x \xi^{(1)} \right) \\ &\quad + \frac{1}{P'} \left(\frac{B^2 \sigma}{|\nabla V|^2} - \frac{\left\langle B^2 \sigma / |\nabla V|^2 \right\rangle}{\left\langle B^2 / |\nabla V|^2 \right\rangle} \frac{B^2}{|\nabla V|^2} \right) \left\langle \delta \hat{P} \right\rangle^{(1)}. \end{aligned} \tag{4.76}$$

The kinetic singular layer equation can be derived from the kinetic vorticity equation (4.65) of next order:

$$
\begin{aligned}
&-\omega^2 \frac{\rho_m}{B^2} \frac{|\nabla V|^2}{} \frac{\partial \mu^{(0)}}{\partial x} \\
&= -\chi' \frac{\partial}{\partial \theta}\left(\frac{|\nabla V|^2}{B^2} \frac{\partial v^{(2)}}{\partial x} - v \frac{\mathbf{B}}{B^2} \cdot \nabla \times \frac{\mathbf{B} \times \nabla V}{B^2} - \tau \frac{\mathbf{B}}{B^2} \cdot \nabla \times \frac{\mathbf{B}}{B^2} + \frac{J'}{\chi'} \tau^{(1)} \right) \\
&- v^{(1)} \left(\frac{J'}{P'} - \frac{\chi'}{P'} \sigma \right) \frac{\partial \sigma}{\partial \theta} - \tau^{(1)} \frac{\chi'}{B^2} \frac{\partial \sigma}{\partial \theta} - \Lambda x \frac{|\nabla V|^2}{\Xi B^2} \frac{\partial}{\partial u} \frac{\partial v^{(1)}}{\partial x} + \frac{P'}{\Xi B^2} \frac{\partial \tau^{(1)}}{\partial u} \\
&+ \frac{\nabla V \cdot \nabla \left(P + B^2\right)}{\Xi B^2 |\nabla V|^2} \frac{\partial \tau^{(1)}}{\partial u} - \frac{\chi}{P'}{}' \frac{\partial \sigma}{\partial \theta} \Theta \frac{\partial \tau^{(1)}}{\partial u} + \frac{\chi'}{P'} \frac{\partial \sigma}{\partial \theta} \frac{\partial}{\partial x} \left(\delta \hat{P}^{(2)} - P' \xi^{(2)} \right) \\
&- \omega \frac{3 n_0 T_{\mathrm{i}} |\nabla V|^2}{B \Omega_{\mathrm{i}} \alpha} \mathbf{e}_2 \cdot \left(\boldsymbol{\kappa} + \nabla \ln B - \mathbf{e}_1 \cdot \nabla \mathbf{e}_1 \right) \frac{\partial^3 \xi^{(1)}}{\partial x^3} \\
&+ \frac{3\omega |\nabla V| P_{\mathrm{i}}}{2\alpha B \Omega_{\mathrm{i}}} \left[\left(\mathbf{e}_2 \cdot \nabla \mathbf{e}_1 \right) \cdot \mathbf{e}_2 \right] \frac{\partial^2 \xi^{(1)}}{\partial x^2} \\
&+ \omega \frac{n_0 T_{\mathrm{i}} |\nabla V|^2}{4\alpha B \Omega_{\mathrm{i}}} \mathbf{e}_2 \cdot \left(8 \nabla \ln B + \frac{1}{2} \boldsymbol{\kappa} \right) \frac{\partial^3 \xi^{(1)}}{\partial x^3} + \mathrm{i}\omega \frac{n_0 T_{\mathrm{i}} |\nabla V|^2}{2 B \Omega_{\mathrm{i}}} \left(\mathbf{e}_1 \cdot \nabla \mathbf{e}_2 \right) \cdot \mathbf{e}_1 \frac{\partial^3 \xi^{(1)}}{\partial x^3} \\
&+ \frac{n_0 e_{\mathrm{i}}}{2} \left(\frac{T |\nabla V|}{m_{\rho \mathrm{i}}} \right)^2 \frac{\partial \ln n_0}{\partial \psi} \left(1 + 2\eta_{\mathrm{i}} \right) \left[\frac{|\nabla \psi|}{\Omega_{\mathrm{i}}^3} \mathbf{e}_1 \cdot \left(\mathbf{e}_1 \cdot \nabla \mathbf{e}_2 \right) - \nabla \cdot \left(\mathbf{e}_2 \frac{|\nabla \psi|}{\Omega_{\mathrm{i}}^3} \right) \right] \frac{\partial^3 \xi^{(1)}}{\partial x^3} \\
&- \frac{n_0 e_{\mathrm{i}}}{2\Omega_{\mathrm{i}}^3} \left(\frac{T |\nabla V|}{m_{\rho \mathrm{i}}} \right)^2 \frac{\partial \ln n_0}{\partial \psi} \left(1 + 2\eta_{\mathrm{i}} \right) \left(\mathbf{e}_1 \cdot \nabla |\nabla \psi| \right) \frac{\partial^3 \xi^{(1)}}{\partial x^3} \\
&+ \mathrm{i}\alpha \frac{n_0 e_{\mathrm{i}} |\nabla V|}{2\Omega_{\mathrm{i}}^3} \left(\frac{T}{m_{\rho \mathrm{i}}} \right)^2 |\nabla \psi| \left[\left(\frac{\partial \ln n_0}{\partial \psi} \right)^2 \left(1 + \frac{15}{2} \eta_{\mathrm{i}}^2 + 4\eta_{\mathrm{i}} \right) + \frac{\partial^2 \ln n_0}{\partial \psi^2} \left(1 + 2\eta_{\mathrm{i}} \right) \right. \\
&\left. + \frac{\partial \ln n_0}{\partial \psi} \left(2 \frac{\partial \eta_{\mathrm{i}}}{\partial \psi} - \eta_{\mathrm{i}} \frac{7}{2} \frac{\partial \ln T}{\partial \psi} \right) \right] \frac{\partial^2 \xi^{(1)}}{\partial x^2}. \qquad (4.77)
\end{aligned}
$$

The simplification of (4.77) is similar to that in section 2.3. It is trivial to obtain $\mu^{(0)}$ from (4.73) and (4.74), and $\tau^{(1)}$ from (4.75). The second, third and eighth terms can be simplified, noting that

$$\left\langle -\nu^{(1)}\left(\frac{J'}{P'}-\frac{\chi'}{P'}\sigma\right)\frac{\partial\sigma}{\partial\theta}-\tau^{(1)}\frac{\chi'}{B^2}\frac{\partial\sigma}{\partial\theta}-\frac{\chi'}{P'}\frac{\partial\sigma}{\partial\theta}\frac{\partial}{\partial x}P'\xi^{(2)}\right\rangle=-\left\langle\sigma\frac{1}{\Xi}\frac{\partial}{\partial u}\tilde{\nu}^{(1)}\right\rangle.$$

The unknown $\nu^{(1)}$ here and in (4.77) can be obtained from (4.76). Using these results, the surface average of (4.77) yields

$$\left(T_0+\omega T_1\right)\frac{\mathrm{d}^3\xi^{(1)}}{\mathrm{d}x^3}+\frac{\mathrm{d}}{\mathrm{d}x}\left(S_0-M_c^*+x^2\right)\frac{\mathrm{d}\xi^{(1)}}{\mathrm{d}x}+\hat{M}_\mathrm{t}\left\langle\frac{\partial\sigma}{\partial\theta}\frac{\mathrm{d}}{\mathrm{d}x}\delta\hat{P}^{(2)}\right\rangle$$

$$+\left(\frac{1}{4}+D_\mathrm{I}\right)\xi^{(1)}-\frac{1}{P'}Gx\frac{\partial}{\partial x}\left\langle\delta\hat{P}\right\rangle^{(1)}-\frac{1}{P'}\left(\frac{1}{4}+D_\mathrm{I}+G\right)\left\langle\delta\hat{P}\right\rangle^{(1)}=0, \quad (4.78)$$

where D_I has been defined in section 2.3, and

$$G=\frac{1}{\Lambda}\left\langle\frac{B^2}{|\nabla V|^2}\right\rangle\left(\langle\sigma\rangle-\frac{\left\langle B^2\sigma/|\nabla V|^2\right\rangle}{\left\langle B^2/|\nabla V|^2\right\rangle}\right),$$

$$M_c^*=\omega(\omega-\omega_{*\mathrm{p}})\frac{N_\mathrm{i}M_\mathrm{i}}{\alpha^2\Lambda^2}\left\langle\frac{B^2}{|\nabla V|^2}\right\rangle\left\langle\frac{|\nabla V|^2}{B^2}\right\rangle,$$

$$\hat{M}_\mathrm{t}=\mathrm{i}\frac{\chi'\Gamma}{\alpha\Lambda^2P'\omega^2}\left\langle\frac{|\nabla V|^2}{B^2}\right\rangle,$$

$$T_0=\frac{n_0e_\mathrm{i}}{2}\left(\frac{T}{m_{\rho\mathrm{i}}}\right)^2\frac{\partial\ln n_0}{\partial\psi}(1+2\eta_\mathrm{i})\frac{\left\langle B^2/|\nabla V|^2\right\rangle}{\alpha\Lambda^2}\left\langle\frac{|\nabla\psi|}{\Omega_\mathrm{i}^3}\mathbf{e}_1\cdot(\mathbf{e}_1\cdot\nabla\mathbf{e}_2)-\nabla\cdot\left(\mathbf{e}_2\frac{|\nabla\psi|}{\Omega_\mathrm{i}^3}\right)\right\rangle,$$

$$T_1=\mathrm{i}\frac{n_0T_\mathrm{i}m_\mathrm{i}}{e_\mathrm{i}\alpha}\frac{\left\langle B^2/|\nabla V|^2\right\rangle}{\Lambda^2}\left\langle\frac{1}{B^2}\mathbf{e}_2\cdot\left[\frac{23}{8}\boldsymbol{\kappa}+\nabla\ln B+\mathrm{i}\frac{1}{2}(\mathbf{e}_1\cdot\nabla\mathbf{e}_1)\right]\right\rangle,$$

$$S_0=-\mathrm{i}\frac{n_0m_{\rho\mathrm{i}}T_\mathrm{i}^2}{2e_\mathrm{i}^2}\frac{\partial\ln n_0}{\partial\psi}(1+2\eta_\mathrm{i})\frac{\left\langle B^2/|\nabla V|^2\right\rangle}{\alpha\Lambda^2}\left\langle\frac{\mathbf{e}_1\cdot\nabla|\nabla\psi|}{B^3}\right\rangle$$

$$+\mathrm{i}\alpha\frac{n_0m_{\rho\mathrm{i}}T_\mathrm{i}^2}{2e_\mathrm{i}^2}\left\langle\frac{|\nabla\psi|}{B^3}\right\rangle\left[\left(\frac{\partial\ln n_0}{\partial\psi}\right)^2\left(1+\frac{15}{2}\eta_\mathrm{i}^2+4\eta_\mathrm{i}\right)+\frac{\partial^2\ln n_0}{\partial\psi^2}(1+2\eta_\mathrm{i})\right.$$

$$\left.+\frac{\partial\ln n_0}{\partial\psi}\left(2\frac{\partial\eta_\mathrm{i}}{\partial\psi}-\eta_\mathrm{i}\frac{7}{2}\frac{\partial\ln T}{\partial\psi}\right)\right].$$

The non-convective pressure $\delta\hat{P}$ in (4.67) can be calculated by the gyrophase-averaged gyrokinetic equation (4.61). We introduce further ordering analyses to reduce the gyrophase-averaged gyrokinetic equation (4.61). The fifth term on the right-hand side of (4.61) is neglected, since it is of order $(R\beta/a)(\lambda_\perp/\lambda_\wedge)$ compared to the first two terms. In the simplification of the last two terms we note that in the singular layer the parallel derivative is reduced by one order, i.e. $\mathbf{e}_b(\mathbf{x})\cdot\nabla_X\xi \sim (\lambda_\perp/R^2)\xi$. Since beta is usually low $\beta \ll 1$ for tokamak confinement, the modification of (2.43) by the term $\langle\delta\hat{P}\rangle^{(1)}$ in (4.75) is small. Therefore, the suppression of compressional Alfvén modes as required by (4.75) leads to $\nabla\cdot\boldsymbol{\xi} = -2\boldsymbol{\kappa}\cdot\boldsymbol{\xi}$. Taking into consideration these simplifications, the gyrophase-averaged gyrokinetic equations for the ion and electron species become

$$
\begin{aligned}
&\left(\mathbf{v}_\parallel\cdot\nabla - \mathrm{i}\omega + \mathrm{i}\omega_\mathrm{d}\right)\delta G_0(\mathbf{X})\\
&= \mathrm{i}\omega\frac{\partial F_{g0}}{\partial\varepsilon}\left(\mu_0 B + v_\parallel^2\right)\left(\boldsymbol{\kappa}\cdot\frac{\mathbf{e}_b\times\nabla V}{B}\mu + \boldsymbol{\kappa}\cdot\frac{\nabla V}{|\nabla V|^2}\xi\right)\\
&\quad + \omega_*^T\frac{1}{\alpha}\frac{\partial F_{g0}}{\partial\varepsilon}\left(\mu_0 B + v_\parallel^2\right)\boldsymbol{\kappa}\cdot\frac{\mathbf{e}_b\times\nabla V}{B}\frac{\partial\xi}{\partial x} + \mathrm{i}\left(\omega - \omega_*^T\right)\frac{e}{m_\rho}\frac{\partial F_{g0}}{\partial\varepsilon}\delta\phi\\
&\quad + 2\frac{\partial F_0}{\partial V}\frac{v_\parallel v_\perp^2}{\Omega^2}|\nabla V|\left(\mathbf{e}_b\cdot\nabla|\nabla V|\right)\frac{\partial^2\xi}{\partial x^2}.
\end{aligned}
\tag{4.79}
$$

Note here that for electrons the last term can be neglected, while for ions this term is of order $(\lambda_\wedge/\lambda_\perp)(L_\mathrm{p}/L_B)(\omega_{*\mathrm{i}}/\omega)$.

To calculate $\langle\delta\hat{P}\rangle^{(1)}$, one needs to calculate $\langle G_0\rangle_\mathrm{b}$, where $\langle\cdots\rangle_\mathrm{b} = \oint(\cdots)\mathrm{d}l/|v_\parallel|$ represents the bounce average. Note that

$$
\frac{m_\rho\Psi_V'}{e}\left(\mu_0 B + v_\parallel^2\right)\left(\boldsymbol{\kappa}\cdot\frac{\mathbf{e}_b\times\nabla V}{B}\right) = \mathbf{v}_\mathrm{d}\cdot\nabla\Psi = v_\parallel\mathbf{e}_b\cdot\nabla\left(\frac{v_\parallel m_\rho g}{eB}\right). \tag{4.80}
$$

The leading order contributions from $\mu^{(0)}$ and $\mathrm{d}\xi^{(1)}/\mathrm{d}x$ in (4.79) vanish after the bounce average. One therefore needs the next order contribution $\tilde{\mu}^{(1)}$ and $\mathrm{d}\tilde{\xi}^{(2)}/\mathrm{d}x$. Note that $\mathrm{d}\tilde{\xi}^{(2)}/\mathrm{d}x$ has already been derived in (4.72), in which $\nu^{(1)}$ can be expressed by (4.76) and $\tau^{(1)}$ by (4.75). As for $\tilde{\mu}^{(1)}$, the dot product of equation $\delta\mathbf{B} = \nabla\times\boldsymbol{\xi}\times\mathbf{B}$ with $\mathbf{B}\times\nabla V/|\nabla V|^2$ yields

$$
\chi'\frac{\partial\tilde{\mu}^{(1)}}{\partial\theta} = \nu^{(1)} - \mathrm{i}\alpha\Lambda x\mu^{(0)} - \frac{(\mathbf{B}\times\nabla V)\cdot\nabla\times(\mathbf{B}\times\nabla V)}{|\nabla V|^4}\xi^{(1)},
$$

where (2.42) has been used. Here, $\mu^{(0)}$ is related to $\xi^{(1)}$ by (4.74) and $\nu^{(1)}$ by (4.76). With $\tilde{\mu}^{(1)}$ and $\mathrm{d}\tilde{\xi}^{(2)}/\mathrm{d}x$ determined, the bounce average of the first two terms on the right-hand side of (4.79) can be calculated. For the final term on the right, since the bounce average in the leading order does not vanish, one does not need to go to the next order.

The effect of $\widetilde{\delta \hat{P}^{(2)}}$ has the same origin as the apparent mass effect M_t in section 2.3. Therefore, only $\mu^{(0)}$ and $d\xi^{(1)}/dx$ are required for evaluating it in (4.79). Noting that $\mu^{(0)}$ is related to $\xi^{(1)}$ by (4.74), $\widetilde{\delta \hat{P}^{(2)}}$ can be fully determined from (4.79).

With $\langle \delta \hat{P} \rangle^{(1)}$ and $\widetilde{\delta \hat{P}^{(2)}}$ determined from the gyrokinetic equation (4.79), the kinetic singular layer equation in (4.78) is fully determined. Equation (4.78) can be used to study the FLR effects on the interchange modes. The kinetic singular layer equation in (4.78) has a similar structure to the two-fluid singular layer equation in [15], which is based on the Braginskii two-fluid equations with the gyroviscous tensor taken into account [7]. The third order derivative term in the singular layer equation is of order $(\lambda_\wedge/\lambda_\perp)(L_p/L_B)$ as compared to the inertia term with the diamagnetic frequency shift. This is not small. The FLR effects of the type $\omega^2 \to \omega(\omega - \omega_{*p})$ can only be obtained in the cylinder model. The toroidal curvature makes the FLR effects asymmetric across the singular surface. This gives rise to the third order derivative in the singular layer equation, (4.78).

4.6 The kinetic ballooning mode theory

In this section we describe the kinetic ballooning mode (KBM) theory. The ideal MHD ballooning theory and general one-dimensional ballooning formalism—the ballooning mode representation—have been described in section 2.4. Therefore, the focus of this section is on the kinetic modification.

The starting equations to describe the KBMs are the gyrokinetic vorticity equation (4.65) and the gyrophase-averaged gyrokinetic equation (4.61). The derivation of the KBM equation is similar to the derivation to the ideal MHD equation in (2.79). Note that the ideal MHD equations are recovered in the gyrokinetic formalism. The suppression of compressional Alfvén modes as required in the ideal MHD case allows the introduction of the so-called stream function $\delta\varphi$, such that $\boldsymbol{\xi}_\perp = \mathbf{B} \times \nabla\delta\varphi/B^2$ [16]. With the ballooning mode representation described in section 2.4, we can proceed to derive the ballooning mode equation. It is convenient to use the so-called Celbsch coordinates (ψ, β, θ) to construct the equations. In these coordinates $\nabla_\perp \to -in\nabla\beta$ and $\mathbf{B} \cdot \nabla = \chi'(\partial/\partial\theta)$. Applying the ballooning eikonal representation (2.78) to the kinetic vorticity equation (4.65), one obtains

$$
\begin{aligned}
&\chi' \frac{\partial}{\partial\theta}\left(|\nabla\beta|^2 \chi' \frac{\partial}{\partial\theta}\delta\varphi\right) + P'\nabla \times \frac{\mathbf{B}}{B^2} \cdot \nabla\beta\delta\varphi + \frac{\omega(\omega - \omega_{*p})}{\omega_A^2}|\nabla\beta|^2\delta\varphi \\
&+ \left(F_2\,|\mathbf{e}_1 \cdot \nabla\beta|^2 + F_3|\mathbf{e}_1 \cdot \nabla\beta|^3\right)\delta\varphi \\
&+ \frac{i}{n}\nabla \times \frac{\mathbf{B}}{B^2} \cdot \nabla\beta\left(\sum\nolimits_{i,e} \frac{m_\rho}{2}\int d^3v v_\perp^2 \delta G_0(\mathbf{x})\right) = 0,
\end{aligned} \tag{4.81}
$$

where F_2 and F_3 describe the FLR modifications additional to the $\omega^2 \to \omega(\omega - \omega_{*p})$ modification of the inertia effect (the third term), which are given, respectively, as follows

$$F_2 = -\frac{3\omega\omega_{*p}}{2\omega_A^2} + n^2\frac{n_0 m_i}{2\Omega_i^2}\left(\frac{T}{m_{\rho i}}\right)^2 \frac{1}{|\nabla\chi|^2}\left\{\frac{\partial \ln n_0}{\partial\chi}(1 + 2\eta_i)(\mathbf{e}_1\cdot\nabla|\nabla\chi|)\right.$$

$$- |\nabla\chi|^2\left[\left(\frac{\partial \ln n_0}{\partial\chi}\right)^2\left(1 + \frac{15}{2}\eta_i^2 + 4\eta_i\right) + \frac{\partial^2 \ln n_0}{\partial\chi^2}(1 + 2\eta_i)\right.$$

$$\left.\left. + \frac{\partial \ln n_0}{\partial\chi}\left(2\frac{\partial\eta_i}{\partial\chi} - \eta_i\frac{7}{2}\frac{\partial \ln T}{\partial\chi}\right)\right]\right\},$$

$$F_3 = n^2\frac{n_0 e_i}{2}\left(\frac{T}{m_{\rho i}}\right)^2\frac{1}{|\nabla\chi|}\frac{\partial \ln n_0}{\partial\chi}(1 + 2\eta_i)\left[\frac{|\nabla\chi|}{\Omega_i^3}\mathbf{e}_1\cdot(\mathbf{e}_1\cdot\nabla\mathbf{e}_2) - \nabla\cdot\left(\mathbf{e}_2\frac{|\nabla\chi|}{\Omega_i^3}\right)\right].$$

One can also apply the ballooning mode representation and transform to the stream function in the gyrokinetic equation, (4.61), to obtain

$$(\mathbf{v}_\parallel\cdot\nabla - \mathrm{i}\omega + \mathrm{i}\omega_\mathrm{d})\delta G_0(\mathbf{X})$$

$$= -n\omega\frac{\partial F_{g0}}{\partial\varepsilon}\left(\mu_0 B + v_\parallel^2\right)\frac{\mathbf{B}\times\boldsymbol{\kappa}}{B^2}\cdot\nabla\beta\delta\varphi + n\omega_*^T\frac{\mathbf{B}\times\left(\mu_0 B\nabla\ln B + v_\parallel^2\boldsymbol{\kappa}\right)}{B^2}\cdot\nabla\beta\delta\varphi$$

$$- \mathrm{i}2n^3\frac{\partial F_0}{\partial\chi}\frac{v_\parallel v_\perp^2}{\Omega^2}(\mathbf{e}_1\cdot\nabla\beta)\left[\mathbf{e}_1\cdot(\mathbf{e}_b\cdot\nabla)\nabla\beta\right]\delta\varphi - \mathrm{i}n^3\frac{\partial F_0}{\partial\chi}\frac{v_\parallel v_\perp^2}{\Omega^2}(\mathbf{e}_1\cdot\nabla\beta)^2(\mathbf{e}_b\cdot\nabla)\delta\varphi$$

$$+ \mathrm{i}\left(\omega - \omega_*^T\right)\frac{e}{m_\rho}\frac{\partial F_{g0}}{\partial\varepsilon}\delta\varphi - \frac{v_\perp^2}{\Omega}\mathbf{e}_1\cdot\nabla_X F_{g0}\mathbf{e}_2\cdot\nabla\left(\frac{1}{B}\mathbf{e}_b\cdot\delta\mathbf{B}\right), \tag{4.82}$$

where $\omega_*^T = -n(P/n_0 e)(\partial \ln F_0/\partial\chi)$. This equation can be used for ions and electrons. As usual, for electrons the FLR terms can be neglected.

The electrostatic potential in (4.82) can be determined by the quasi-neutrality condition, (4.66),

$$\delta\phi(\mathbf{x}) = -\frac{1}{\sum_{\mathrm{i,e}} n_0 e^2/T}\sum_{\mathrm{i,e}} e\int \mathrm{d}^3v\delta G_0. \tag{4.83}$$

The perturbed parallel magnetic field in (4.82) can be determined by $\mathbf{e}_b\cdot\delta\mathbf{B} = \mathrm{i}((|\nabla\chi|/nB)\mathbf{e}_1\cdot\nabla\times\delta\mathbf{B}$. To prove this relation the high-n mode assumption has been used. This procedure to determine $\delta B_\parallel$ is the same as in [13]. In the current approach the radial perturbed current density $\mathbf{e}_1\cdot\nabla\times\delta\mathbf{B}$ has been evaluated in (4.63). Note that the calculation of another component $\mathbf{e}_2\cdot\nabla\times\delta\mathbf{B}$ in (4.64) is similar to that of

$\mathbf{e}_1 \cdot \nabla \times \delta\mathbf{B}$. This again shows that the current approach does not lead to much further complexity compared to the approach in [13] by deriving the vorticity equation from the averaged gyrokinetic equation. The current approach avoids the difficulty of transforming back the vorticity equation to the configuration space. The FLR effects on the ballooning modes in [13] are over-simplified, mainly because this back transform has not been made. To save space, and also considering that in many ordering schemes this term is unimportant, we do not detail the expression for $\mathbf{e}_1 \cdot \nabla \times \delta\mathbf{B}$ in the ballooning formulation. Readers can obtain it in a straightforward manner.

Equations (4.81), (4.82), (4.83) and (4.63) constitute a complete set of equations for determining the linear kinetic ballooning stability. Note that the quasi-neutrality condition (4.83) needs to be inserted into the gyrokinetic equation (4.81) and the gyrokinetic equation is the kinetic 'replicate' of the parallel momentum equation. This indicates that the sound wave coupling to the shear Alfvén mode, in particular the sound wave spectrum as shown for example in figure 2.1, is modified by the electrostatic waves [8].

4.7 Discussion

In this chapter we have described the gyrokinetic theory. This includes the early efforts both in the electrostatic [1, 2] and electromagnetic [3, 4] gyrokinetic theories, and also the recent revisiting of this theory [5]. The theory is then applied to study both the electrostatic and electromagnetic modes.

For electrostatic drift waves, the ITG modes have been studied. The emphasis is placed on investigating the eigenmode structure and the free boundary problem. Since the H-mode edge contains the non-neutral charges, the current theory can be extended to study their effects. We also note that electrostatic modes can change the sound wave spectrum of electromagnetic modes. This shows that the electrostatic modes can still be meaningful in the finite beta case. The parallel electrostatic wave can be excited along the total (equilibrium and perturbed) magnetic field.

For electromagnetic modes, both the singular layer and ballooning modes are investigated. It is shown that the FLR effects are much more complicated than previously thought in this field. The inclusion of an equilibrium distribution function to sufficient order and the retention of the gyrophase-dependent contributions in the gyrokinetic formalism make the difference. The KBM formalism developed in section 4.6 can also be used to study the kinetic TAE modes.

Bibliography

[1] Rutherford P H and Frieman E A 1968 Drift instabilities in general magnetic field configurations *Phys. Fluids* **11** 569–85

[2] Taylor J B and Hastie R J 1968 Stability of general plasma equilibria: I. formal theory *Plasma Phys.* **10** 479

[3] Antonsen T M and Lane B 1980 Kinetic equations for low frequency instabilities in inhomogeneous plasmas *Phys. Fluids* **23** 1205–14

[4] Catto P J, Tang W M and Baldwin D E 1981 Generalized gyrokinetics *Plasma Phys.* **23** 639

[5] Zheng L J, Kotschenreuther M T and Van Dam J W 2007 Revisiting linear gyrokinetics to recover ideal magnetohydrodynamics and missing finite Larmor radius effects *Phys. Plasmas* **14** 072505

[6] Hazeltine R D 1976 Review of neoclassical transport theory *Advances in Plasma Physics* vol 6 ed A Simon and W B Thompson (New York: Wiley) p 273

[7] Braginskii S I 1966 Transport processes in plasma *Reviews of Plasma Physics* vol 1 ed M Leontovich (New York: Consultants Bureau) pp 205–311

[8] Zheng L-J and Tessarotto M 1994 Collisionless kinetic ballooning equations in the comparable frequency regime *Phys. Plasmas* **1** 2956–62

[9] Romanelli F, Chen L and Briguglio S 1991 Kinetic theory of the ion-temperature-gradient-driven mode in the long wavelength limit *Phys. Fluids* B **3** 2496–505

[10] Abramowitz M and Stegun I A ed 1972 *Handbook of Mathematical Functions* (Washington, DC: National Bureau of Standards)

[11] Chen L and Tsai S 1983 Electrostatic waves in general magnetic field configurations *Phys. Fluids* **26** 141–5

[12] Antonsen T M and Lee Y C 1982 Electrostatic modification of variational principles for anisotropic plasmas *Phys. Fluids* **25** 132–42

[13] Tang W, Connor J and Hastie R 1980 Kinetic-ballooning-mode theory in general geometry *Nucl. Fusion* **20** 1439

[14] Glasser A H, Greene J M and Johnson J L 1975 Resistive instabilities in general toroidal plasma configurations *Phys. Fluids* **18** 875–88

[15] Zheng L-J 1993 Two fluid equations for low *n* singular modes in the low frequency regime *Phys. Fluids* B **5** 1962–70

[16] Chance M S *et al* 1979 MHD stability limits on high-beta tokamaks *Proc. 7th Int. Conf. Plasma Physics and Controlled Fusion Research (Innsbruck, Austria, 23–30 August, 1978)* vol 1 (Vienna: International Atomic Energy Agency) p 677

Chapter 5

Physical interpretations of experimental observations

In this chapter we discuss the physical interpretations of experimental observations. We have described the main topics in tokamak stability theories in previous chapters. Using these theories one can construct intuitive physics pictures for analyzing tokamak experiments. Both core and edge stability and transport phenomena are discussed in this chapter.

The magnetic confinement of fusion plasmas is a challenging topic. Nevertheless, its physics can still be quite comprehensible. This is largely because there is a strong magnetic field in the system. We first note that one of the important features of the tokamak concept lies in its toroidal symmetry. From classical mechanics we know that toroidal symmetry allows the existence of closed magnetic flux surfaces and the conservation of toroidal canonical angular momentum. These general properties help in tokamak confinement. However, to maintain toroidal symmetry one needs to induce a toroidal current. Due to the current interchange phenomenon the interchange-type modes in tokamaks tend to become tearing modes, which break the symmetry and the perfect magnetic surfaces. For high-n modes this enhances transport due to the appearance of island and stochastic regions, which may overlap. For the low-n case this can result in the conversion of kink modes to tearing modes, which may cause global disruptions. This leads to the tokamak confinement becoming worse than one would expect for a symmetric system and worse than one would expect from the kink mode picture. Nonetheless, Ohmic heating can be used for profile control, pinching the plasma column to achieve better confinement.

The interaction of the toroidal current with the externally applied magnetic field can also be exploited to improve tokamak confinement. In section 1.3 we have shown that the vertical magnetic field can be used to control the magnetic surface shift, while in section 2.3 we have shown that the Mercier criterion changes. This concept can also be used to control the internal kink modes, etc. In addition, we know that the L- and H-modes' power thresholds depend on the divertor positions.

doi:10.1088/978-1-6270-5423-2ch5

Controlling the divertor positions should also help to suppress edge localized modes (ELMs). These examples show that fundamental physical considerations can lead to a great scope for improvement of tokamak confinements.

In the following sections we will discuss the physical interpretations for various experimentally observed phenomena.

5.1 The tokamak confinement modes

Before discussing the individual experimental phenomena, we first discuss various tokamak confinement modes and provide general classifications.

In this field the peeling–ballooning mode stability boundaries are often used to discuss tokamak edge stability [1]. Recently, the EPED model has been adopted for predicting the H-mode pedestal height and width based upon two fundamental and calculable constraints: (1) the onset of nonlocal peeling–ballooning modes at low to intermediate mode numbers and (2) the onset of nearly local kinetic ballooning modes at high mode numbers [2]. Some theoretical issues need to be discussed here. The first regards the applicability of the ideal MHD model for peeling–ballooning mode calculations. At the H-mode plasma edge the pressure gradient is steep, but the density can be low. This causes the diamagnetic frequency to become very large. Even for $n = 3$ modes the diamagnetic frequency can be comparable to the ion transit frequency at the pedestal [3]. Due to wave–particle resonances and excitation of the parallel electric field the ideal MHD becomes inapplicable in this case. The second theoretical issue concerns the applicability of the localized kinetic ballooning mode calculations. The conventional ballooning mode invariance breaks down at the pedestal region for a reasonably high mode number. One needs a free boundary kinetic code to assess the linear stability of the pedestal. Nonetheless, one does see some success for the EPED model in explaining experimental observations. This indicates the simple nature of tokamak discharges: the pressure gradient and magnetic shear can be used to classify similar tokamak discharges. The simple s–α stability diagram for ballooning–peeling modes as reproduced in figure 2.3 can still provide very informative stability indications. Due to the current interchange effects, as discussed in section 3.3, the birth of the tearing modes is also related to the interchange-type modes. Therefore, the Rayleigh–Taylor instability drive and magnetic shear stabilization remain key players, but not the only ones.

Before we discuss the experimental phenomena individually in the following sections, we provide a general classification of confinement modes. We first point out that a single stability boundary, for example the peeling–ballooning stability boundary, does not seem to naturally fit various confinement modes. In figure 5.1 we use multiple stability boundaries to explain various confinement modes. The stability boundaries can be nonlinearly explosive or nonexplosive instabilities.

One can expect that at low beta there is a stable region, which is related to the so-called L-mode. The edge modes are not critical in this case. If the internal ballooning s–α stability diagram in figure 2.3 is used to delineate it, the L-mode corresponds to the first stability regime at a low α value.

As depicted in figure 5.1, the H-mode stability regime occurs at higher beta. In this regime, the major pressure drop, i.e. the pressure gradient, is accumulated at

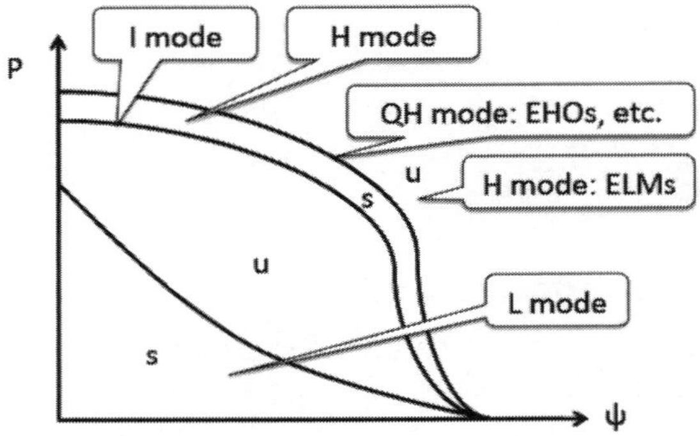

Figure 5.1. A schematic explanation of various tokamak confinement modes, using s and u to denote the stable and unstable regions.

the so-called pedestal region at the plasma edge, while at the large core region the pressure profile remains flat, or the pressure gradient there is small. The ballooning s–α stability diagram in figure 2.3 can, again, be used to explain this. The core region belongs to the first stability regime because the α value is small due to the low pressure gradient. The pedestal region, instead, corresponds to the second stability regime. In the conventional ballooning mode theory the local magnetic shear induced by the Pfisch–Schlüter current plays a key role in the appearance of the second stability regime. The large pressure gradient leads to a large Shafranov shift, thus the local magnetic shear increases and the modes are stabilized. The pedestal stability, however, is attributable to more factors. Because of the ion loss induced by the finite banana orbit size, accumulation of the excessive electrons at the plasma edge occurs. This negative charge accumulation at the edge results in a strongly sheared plasma rotation. Both the rotation and the non-neutral electron shell can help the pedestal maintain linear stability. In addition, the low density and high pressure gradient can also result in a large bootstrap current at the pedestal. As will be discussed in section 5.3 this can help to create a transport barrier at the pedestal. There is clear experimental evidence that shows that pedestal stability is more than just the pedestal height and width. For example, the H-mode power threshold depends on the divertor position and the H-mode confinement is limited by the density limit, i.e. the so-called Greenwald limit [4].

The extra stability factors in the pedestal help in understanding why the H-mode has a density limit. At a fixed pedestal pressure the lower density results in a higher temperature. The direct consequence of a high pedestal temperature is stronger ion loss due to the increased ion orbit size. This helps in the stabilization of the edge rotation and non-neutral shell. In addition, the low density can contribute to a larger bootstrap current. This also benefits the building-up of a edge transport barrier. The dependence of the H-mode threshold on the divertor position also confirms the role of edge excessive charges. The divertor position that causes the greatest number of ion losses favors the achievement of the H-mode.

Just as there is an unstable region between the first and second stability regimes for the ballooning modes in figure 2.3, one can anticipate that an unstable region exists between the L- and H-modes in figure 5.1. High power is needed to push the system from the L- to H-mode before instabilities can develop. This explains the existence of the H-mode power threshold.

In the H-mode stability regime a large amount of free energy is stored in the pedestal region. This causes the tokamak edge to be prone to the excitation of ELMs. As will be discussed in section 5.5, due to the peeling-off phenomenon, the stabilizing excessive charges at the plasma edge tend to become lost. Coupled to the scrape-off layer (SOL) current, the ELMs tend to be virulent.

Both the I-mode and quiescent H-mode (QH-mode) are operating schemes for avoiding ELMs in H-mode confinement. As shown in figure 5.1, the I-mode pushes the system to the lower stability boundary, while the QH-mode tries to maintain the system at the upper stability boundary. Control of the heating power or the excitation of non-damaging internal modes can be used to reach these modes. This issue will be discussed further in section 5.7.

5.2 Enhanced electron transport

We first discuss enhanced electron transport in tokamaks. In tokamak experiments it is often observed that electron transport is stronger than ion transport. In the classical transport picture, for a system with perfect magnetic surfaces electron transport should be lower than ion transport since the electron has a much smaller Larmor radius than the ion. This anomaly has led to the consideration of turbulence transport, related in particular to the modes that resonate with electrons, for example, the electron temperature gradient modes and trapped electron modes.

However, from the experimentally observed turbulence frequency spectrum one can find that the modes in the frequency domain comparable with the ion transit frequency persist. In many cases this spectrum is stronger than that for the higher frequency modes. The turbulences with ion transit frequencies should lead to anomalous ion transport in the same way. Therefore, anomalous transport alone cannot fully explain why only the electron transport is enhanced.

In [5], broken magnetic surfaces due to the formation of magnetic islands and stochastic field lines are used to explain the enhanced electron transport. However, the authors do not explain the formation of magnetic islands and stochastic field line regions in axisymmetric tokamak plasmas. This is explained in [6] using the current interchange modes. The current interchange phenomenon, as described in section 3.3, causes interchange-type instabilities, including various drift waves, to convert to current interchange tearing modes, see figure 3.2. This helps to clarify the source of electron transport in tokamaks. Since electrons have large parallel thermal velocities, electron transport can be dramatically enhanced by broken magnetic surfaces.

The formation of magnetic islands and stochastic field lines can also enhance impurity transport.

5.3 Transport barriers

In this section we discuss the transport barriers observed in tokamak experiments. Both turbulence and field line stochastic transport are discussed.

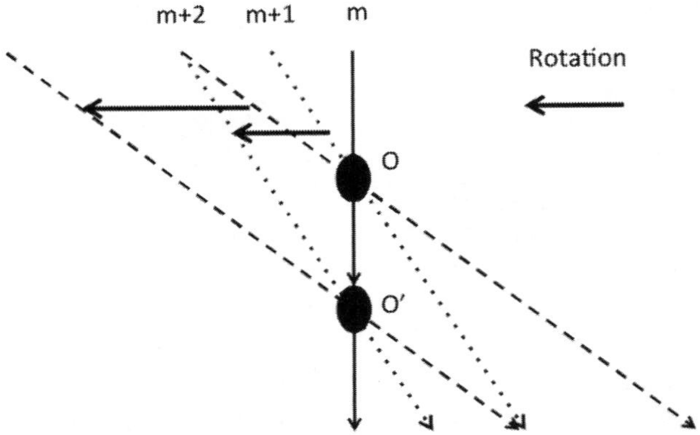

Figure 5.2. A schematic explanation of why the flow shear does not decorrelate turbulence eddies in a system with magnetic shear.

For turbulence transport the conventional explanation is based on the so-called flow shear decorrelation. In fact, this picture is not generally correct for systems with magnetic shear. We use figure 5.2 to explain this [7]. In figure 5.2, the solid, dotted and dashed long arrows represent, respectively, the magnetic field lines on the $q = m/n$, $(m + 1)/n$ and $(m + 2)/n$ surfaces. We use the rotating frame of the m/n surface as the reference frame. It is assumed that only the rotation shear exists, but no rotation curvature. Therefore, when the magnetic field line on the $(m + 1)/n$ surface moves one step left in the time interval Δt, the magnetic field line on the $(m + 2)/n$ surface moves two steps in the same time interval Δt. If the original turbulence eddy is located at point 'O' and after the time interval Δt the eddy moves to point 'O'', from figure 5.2 one can see that the turbulence eddy is not decorrelated. This indicates that the flow shear does not decorrelate turbulence eddies. The observation of the decorrelation of turbulence eddies in the numerical simulation is simply due to the fact that the observation is not made in the moving frame. The eddies propagate along the field lines helically around the plasma torus. If the simulation runs for long enough, one should see the recurrence of the eddies. Only flow curvature can result in turbulence decorrelation in a system with magnetic shear. This resembles the ballooning mode behavior in rotating plasmas with the Cooper representation [8]. Although the Cooper representation does not represent a linear eigenmode, the turbulence decorrelation actually involves nonlinear mode coupling. The detailed mode and field line patterns need to be considered.

From the analyses based on figure 5.2 one can see that decorrelation can indeed occur when the magnetic shear vanishes or is very small. This coincides with the fact that the transport barrier is located at the extremes of the safety factor. The internal transport barrier occurs at the q minimum. At the pedestal the strong bootstrap current can also cause the magnetic shear to reverse or to be reduced significantly [9].

In addition to turbulence transport, we also need to discuss the field line stochastic transport. In particular, we note that the unstable modes underlying the turbulence transport can convert to current interchange tearing modes. As shown in figure 5.3,

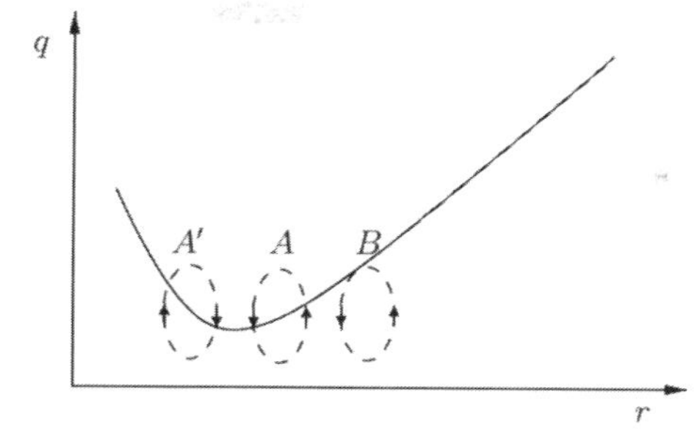

Figure 5.3. A schematic explanation of the formation of the so-called ghost surface at the q minimum. (Reproduced with permission from [6]. Copyright 2010, AIP Publishing LLC.)

the island rotational transform reverses direction across a q minimum. As islands grow and reconnect in the case without magnetic shear reversal (islands 'A' and 'B'), magnetic energy can be released and the radial transport step size is increased. However, this type of island–island reconnection is prohibited or reduced at the stationary surface of the safety factor in the case of reversed shear (islands 'A' and 'A''). Reconnection forbidden layers have been observed at the q maximum in reversed field pinch experiments [10]. These layers are known as the so-called ghost surfaces. The ghost surfaces were originally introduced for the standard map to denote surfaces which are non-intersecting (in this context, they are called ghost circles) [11, 12]. It should be noted that at the q stationary surface such types of ghost surfaces can be formed naturally due to different wiring directions of the island field lines on the opposite sides of the q stationary surface.

We have discussed the stochastic transport at the internal transport barrier, where q is minimum. In the pedestal region the strong bootstrap current can also cause the safety factor to reverse [9]. A safety factor maximum can be formed in this case. Similarly, a stochastic transport barrier can result.

5.4 Nonlocal transport

In this section we discuss the so-called nonlocal transport observed in tokamak experiments [13]. It has been found that a sudden cooling at the plasma edge can cause the core temperature to rise. It has been shown that Ohmic heating power redistribution is insufficient to explain the temperature increase at the center and the conventional local transport models are unable to explain this phenomenon. We point out that Ohmic current redistribution can also lead to compression of the center plasma. The compression can cause a stronger Ohmic power coupling and therefore provide a significant additional heating mechanism for the center plasma.

In figure 5.4 we use the circuit picture to explain the role of Ohmic heating power coupling in the nonlocal transport experiments. The tokamak plasma torus is an open system, not only to the wall but also to the coupling of Ohmic power. Keeping the total current constant, cooling at the edge plasma in the experiments causes the loop voltage to increase. The power from the Ohmic heating from the other side of the transformer is determined by IV_ϕ, i.e. the product of total current and loop voltage.

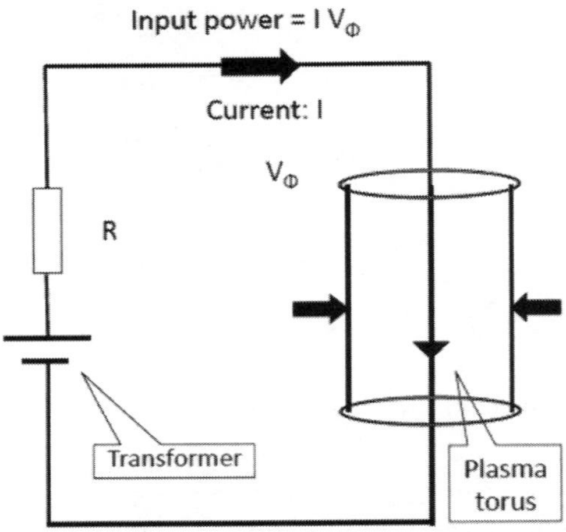

Figure 5.4. A circuit diagram to explain nonlocal transport.

A direct consequence of cooling the plasma edge while keeping the total current constant, is the increase of loop voltage and consequently Ohmic heating power.

Note that due to the cooling at the edge the Ohmic heating power tends to deposit at the core plasma. This causes the center current to increase. The increase of the center current can cause the plasma to compress due to the Z-pinch effect. Note that Ohm's law in this case is

$$J_\phi = \sigma\left(E_\phi + v_r B_\theta\right),$$

where the symbols assume their usual meanings. The compression effect (negative radial velocity v_r) tends to reduce the toroidal current density further. In holding the total current constant, the loop voltage must increase. Therefore, the core plasma is heated and the core current increases further. This causes the plasma to be further compressed. This is a positive feedback process, which helps to explain the nonlocal transport observed experimentally.

Similarly, we point out that off-axis heating can produce an Ohmic current peak locally. This type of hollow current profile can effectively cause the plasma column to compress. Due to the localization of the current peak, the compression tends to localize around the peak. Consequently, the plasma is swept inward and the local pressure gradient builds up. That is why suitable off-axis heating can create a pressure profile as if there was a so-called internal transport barrier. Similarly, off-axis heating can considerably reduce the H-mode power threshold.

Note that a tokamak can be a Z-pinch. Suitable control of the heating sequence and deposition position can significantly modify the targeting plasma confinement properties.

5.5 The edge localized modes

In this section we discuss the ELMs. H-mode confinement has currently been adopted as a reference for next generation tokamaks, especially for ITER. However,

H-mode confinement is often linked to damaging ELMs [14]. There is concern that ELMs can discharge particles and heat into the SOL and subsequently to the divertors. The divertor plates can potentially be damaged by such a discharge.

The most well-known theories for explaining the ELMs are the peeling and peeling–ballooning modes [15, 1]. However, the peeling or peeling–ballooning modes are of the kink or pressure driven types. Without field line reconnection the plasmas inside the last closed flux surface are in fact not peeled off. Note that there is a current jump across the last closed flux surface between the edge plasma and the SOL. As shown in section 3.3 the existence of a current gradient can cause the interchange-type modes to convert to tearing modes. The current jump at the plasma edge indicates that a very large current gradient exists there. As shown in [16], the current jump can cause the external kink modes to readily convert to tearing modes. In particular the peeling and peeling–ballooning modes can convert the peeling-off modes. Magnetic reconnection in the presence of tearing modes subsequently causes the tokamak edge plasma to be peeled off to link to the divertors.

We further note that due to the large ion orbit size, there are ion losses at the plasma edge [18]. This causes the plasma edge to usually carry negative charges. In contrast, due to the large electron parallel velocities the divertor sheaths have an excess of positive charges. Field line reconnection between the plasma edge and the SOL can heat up the the SOL and neutralize the positive charges at the divertor sheaths. As a result the SOL current tends to burst. This current burst can further enhance tearing mode activities, leading to a positive feedback process. Using this process for ELM interpretation, many ELM characteristics can be explained naturally [19].

First, the co-occurrence of the D_α (i.e. MHD activities) and the SOL current during the ELM activities can be explained. Indeed, figure 5.5 shows the DIII-D tokamak experimental observation of ELM activities, in which one can see that the MHD modes and SOL current are closely tied to each other.

Second, this picture can also explain the experimental observation that the ELM strength decreases as the pedestal collisionality increases. Note that the excessive electrons at the plasma edge take time to travel to the divertor sheaths. The larger the pedestal collisionality, the longer the travel time. A longer travel time results in a larger delay for the response to the MHD activities. Consequently, a weaker coupling between the SOL current and MHD modes results. Since the ELM bursts rely on a positive feedback process, a weaker coupling leads to a weaker ELM strength.

Pellet injection and the introduction of resonant magnetic perturbation for ELM mitigation can both be interpreted as causing an increase of the connection length between the pedestal and the divertor sheaths.

ELM bursting is a nonlinear and edge localized process. One has to consider the edge peculiarities to understand the physics, such as the edge current jump, the SOL current, the edge electric charges, the divertor sheaths, etc. These features may not be simply specified by the pedestal height and width. Keeping these features in mind, one can understand why there are no ELM type bursts at the internal transport barrier.

The ELMs are a cause of concern regrading divertor damage. However, this does represent a energy release from the tokamak confinement without causing a global disruption. Due to the current interchange effects discussed in section 3.3, the low-n

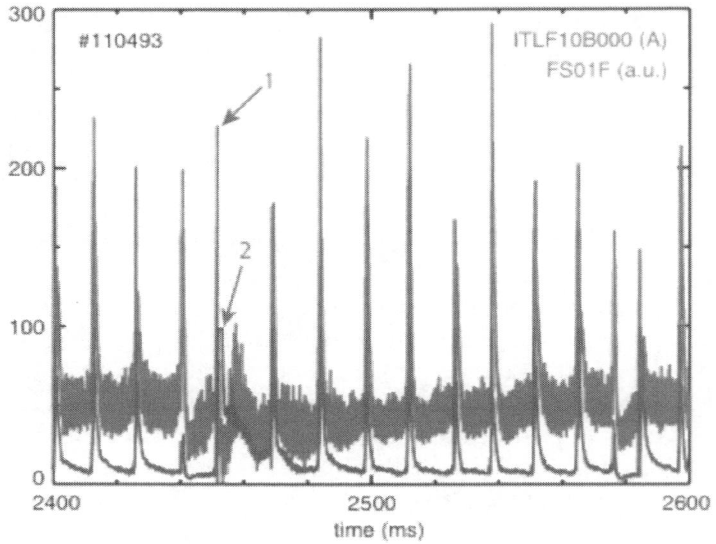

Figure 5.5. The D_α and SOL current bursts observed during ELM activities. (Reproduced with permission from [17]. Copyright 2004, IAEA.)

kink modes in the tokamak can also convert to tearing modes. For the low-n global kink modes, however, the consequence is a major disruption. But, in the ELM case only a small part of the edge plasma is peeled off, the majority of the core plasma remains largely intact. The SOL current in ELM bursts is replaced by the halo current in the disruption. Therefore, in this sense ELMs can somehow be a 'favorable' feature of H-mode confinement, since it is much easier to deal with ELMs than with the disruption. The ELMs are just micro-disruptions which help to release the tokamak's over-stressed energy.

5.6 Blob transport

In this section we discuss the non-one-dimensional transport phenomena at the plasma edge and SOL. Experiments show that the ELMs can develop into filamentary tubes, which align along the magnetic field and have different thermal properties, such as density and temperature, than their ambient plasma [20]. These filamentary tubes move toward the wall convectively and this is referred to as blob transport. In fact, at the plasma edge a similar transport phenomenon, intermittent transport, is also observed.

The conventional anomalous transport is actually a one-dimensional theory in the radial direction. The transport due to the stochastic field is focused on parallel transport. Blob transport is beyond these two theoretical frameworks. For a long time in this field, extended resistive interchange mode theories have been used to interpret this phenomenon [21]. In the SOL the field lines terminate at the divertors. Due to the divertor sheath effects the virtual resistivity can be large. Therefore, the electric field can build up across the blob. Consequently, the $\mathbf{E} \times \mathbf{B}$ velocity pushes the blob toward the wall.

Note that the blob has a different temperature and density from its ambient plasma and that there is loop voltage in the tokamak. Consequently, the blob should carry a different current from its ambient plasma. This causes the blob to carry its own current and have its own poloidal magnetic field, i.e. the blob should be a magnetic island. The rotational transform of the island's magnetic field can cause the charge separation inside the islands to be neutralized. This shows that the conventional $\mathbf{E} \times \mathbf{B}$ drift picture is insufficient to explain blob transport.

Blob transport can be explained naturally from the magnetic island picture. Since the blobs are aligned with the magnetic field lines, they are curved spatially. Consequently, there are hoop forces on them, see figure 1.2 for an example. As discussed in section 1.3, the tokamak hoop forces are balanced by the application of a vertical magnetic field. Since the blobs surround the plasma column helically, the vertical magnetic field cannot provide the balance force. One can imagine that, in this case, the blobs should be pushed toward the wall by the hoop force. In view of the fact that the magnetic field lines are nested in tokamak systems, the forced field line reconnection process should continue to occur during the blob movement toward the wall. In this field, most blob simulations are performed in a straight system. When a bended system is used, one should recover the hoop force effects and readily reproduce the blob transport.

From the ELM picture discussed in the previous section, 5.5, and the current interchange tearing mode theory in section 3.3, one can see that the tearing modes cause the birth of blob or intermittent transport. The transport of fully developed islands as a block—i.e. blob or intermittent transport—driven by the island hoop force is simply a subsequent development of the tearing modes, as soon as further reconnection allows the blob movement.

Experimental observations show that the blobs at the plasma edge are brighter than the surrounding areas. The current interchange tearing mode picture can also explain this phenomenon naturally. From figure 3.2, one can see that islands formed by the current interchange tearing modes bring cold and hot plasmas together. Therefore, one can anticipate that the charge recombination process occurs in the islands. This recombination leads to the brightness. The tearing mode picture can also explain other blob transport features.

5.7 Edge harmonic oscillations

As discussed in section 5.1, QH-modes exists in tokamak discharges. A distinguishing feature of the QH-modes is the excitation of the so-called edge harmonic oscillations (EHOs) or outer modes [22, 23]. In contrast to ELMs, EHOs are mild MHD activities which can pump out plasma energy steadily without damaging the divertors. The excitation of non-damaging modes to pump out energy in order to keep the system near the marginal states is a general measure to avoid ELMs, for example in the I-mode case there are the quasi-coherent modes. As an extreme measure one can simply increase the ELM frequency to reduce the energy released per ELM burst. In this section we simply explain the EHO physics.

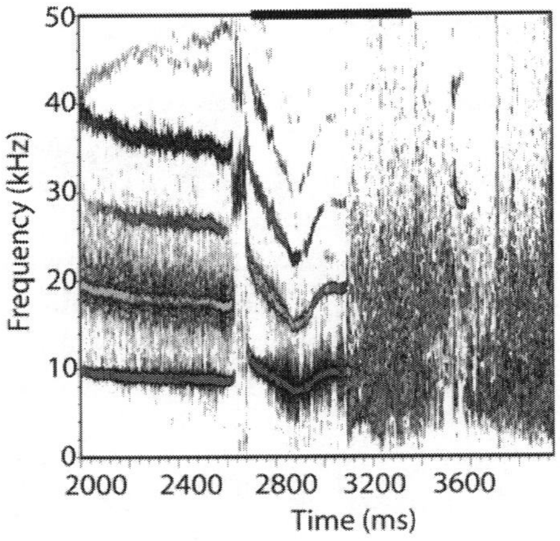

Figure 5.6. Edge harmonic oscillations in QH-mode discharge. (Reproduced with permission from [22]. Copyright 2005, AIP Publishing LLC.)

Figure 5.6 shows the frequency spectrum observed in DIII-D tokamak QH-mode discharge. The EHOs appear during approximately 2000–2600 ms in this figure. Because EHOs pump out the heat, there are no ELMs in this period. The typical EHO frequency features are as follows: for $n = 1$ modes the frequency is the rotation frequency near the pedestal top and for $n > 1$ modes the frequencies are the n-multiple of the $n = 1$ mode frequency.

Note that the QH-mode usually occurs in hot temperature or low density discharges, where the collisionality is low. As is well known, the bootstrap current increases as the collisionality decreases. Therefore, the QH-mode tends to have a strong bootstrap current at the pedestal region. A strong bootstrap current may cause the safety factor profile to reverse or to form a plateau in the pedestal region.

The EHO frequency features and the possible reversal of the safety factor profile tend to suggest that the EHOs are simply the infernal modes in the pedestal region [24]. Since the infernal modes are internal modes, the conversion to peeling-off modes can be avoided and thus the edge negative charge shell can basically be kept intact. Because of the localization of infernal modes, their frequencies are determined by the local resonance harmonics. With rotation present one can expect the infernal modes to have a frequency multiplying role as the EHOs. The infernal modes are of the snaky type, as observed experimentally. In fact, snaky-type modes are observed both around the internal transport barrier, where q has a minimum, and the QH-mode pedestal, where q has a maximum if the bootstrap current is taken into account. The snaky-type modes at the internal transport barrier have also been interpreted as the infernal modes.

When tokamak energy builds up, it must be released in one way or another. A non-damaging process, such as EHOs, can be used to release the over-stressed energy storage to avoid the damaging ELMs. This is the main principle, although the types of non-damaging modes may vary.

Bibliography

[1] Snyder P, Wilson H, Ferron J, Lao L, Leonard A, Mossessian D, Murakami M, Osborne T, Turnbull A and Xu X 2004 ELMs and constraints on the H-mode pedestal: peeling–ballooning stability calculation and comparison with experiment *Nucl. Fusion* **44** 320

[2] Snyder P B, Osborne T H, Burrell K H, Groebner R J, Leonard A W, Nazikian R, Orlov D M, Schmitz O, Wade M R and Wilson H R 2012 The EPED pedestal model and edge localized mode-suppressed regimes: studies of quiescent H-mode and development of a model for edge localized mode suppression via resonant magnetic perturbations *Phys. Plasmas* **19** 056115

[3] Zheng L J, Kotschenreuther M T and Valanju P 2014 Diamagnetic drift effects on the low-*n* magnetohydrodynamic modes at the high mode pedestal with plasma rotation *Phys. Plasmas* **21** 062502

[4] Greenwald M 2002 Density limits in toroidal plasmas *Plasma Phys. Control Fusion* **44** R27

[5] Rechester A and Rosenbluth M 1978 Electron heat transport in a tokamak with destroyed magnetic surfaces *Phys. Rev. Lett.* **40** 38–41

[6] Zheng L J and Furukawa M 2010 Current-interchange tearing modes: conversion of interchange-type modes to tearing modes *Phys. Plasmas* **17** 052508

[7] Zheng L J and Tessarotto M 1997 private communication

[8] Waelbroeck F L and Chen L 1991 Ballooning instabilities in tokamaks with sheared toroidal flows *Phys. Fluids* B **3** 601–10

[9] Kessel C *et al* 2007 Simulation of the hybrid and steady state advanced operating modes in ITER *Nucl. Fusion* **47** 1274

[10] Puiatti M E *et al* 2009 Helical equilibria and magnetic structures in the reversed field pinch and analogies to the tokamak and stellarator *Plasma Phys. Control. Fusion* **51** 124031

[11] Morrison P J and Wurm A 2009 Nontwist maps *Scholarpedia* **4** 3551

[12] del Castillo-Negrete D and Morrison P J 1993 Chaotic transport by Rossby waves in shear flow *Phys. Fluids* A **5** 948–65

[13] Gentle K W *et al* 1997 The evidence for nonlocal transport in the Texas experimental tokamak *Phys. Plasmas* **4** 3599–613

[14] Wagner F *et al* 1982 Regime of improved confinement and high beta in neutral-beam-heated divertor discharges of the ASDEX tokamak *Phys. Rev. Lett.* **49** 1408–12

[15] Wilson H R, Snyder P B, Huysmans G T A and Miller R L 2002 Numerical studies of edge localized instabilities in tokamaks *Phys. Plasmas* **9** 1277–86

[16] Zheng L J and Furukawa M 2014 Peeling-off of the external kink modes at tokamak plasma edge *Phys. Plasmas* **21** 082515

[17] Takahashi H, Fredrickson E, Schaffer M, Austin M, Evans T, Lao L and Watkins J 2004 Observation of SOL current correlated with MHD activity in NBI heated DIII-D tokamak discharges *Nucl. Fusion* **44** 1075

[18] Hazeltine R D, Xiao H and Valanju P M 1993 Gyrosheath near the tokamak edge *Phys. Fluids* B **5** 4011–4

[19] Zheng L J, Takahashi H and Fredrickson E D 2008 Edge-localized modes explained as the amplification of scrape-off-layer current coupling *Phys. Rev. Lett.* **100** 115001

[20] Nedospasov A 1992 Edge turbulence in tokamaks *J. Nucl. Mater.* **196–198** 90–100

[21] Krasheninnikov S 2001 On scrape off layer plasma transport *Phys. Lett.* A **283** 368–70

[22] Burrell K H *et al* 2005 Advances in understanding quiescent H-mode plasmas in DIII-D *Phys. Plasmas* **12** 056121

[23] Solano E R *et al* 2010 Observation of confined current ribbon in JET plasmas *Phys. Rev. Lett.* **104** 185003

[24] Zheng L, Kotschenreuther M and Valanju P 2013 Low-*n* magnetohydrodynamic edge instabilities in quiescent H-mode plasmas with a safety-factor plateau *Nucl. Fusion* **53** 063009

Advanced Tokamak Stability Theory

Linjin Zheng

Chapter 6

Concluding remarks

According to the Bible, when God created the Universe, He claimed ‘it is good’ and ‘it is so’ (Genesis 1). Beauty (the ‘good’) and simplicity (the ‘so’) are the inherent nature of our Universe. The axisymmetry of the tokamak concept contains both the features of beauty and simplicity. In addition, the safety factor can also minimize the poloidal and up–down asymmetries. The symmetry implies there are Hamiltonian invariants in the system. These aspects make the tokamak concept unique. The rules of simplicity and beauty are universal.

Due to the novel intrinsic features of tokamaks, even though our theoretical understanding may be lagging, experiments continue to advance, producing exciting discoveries. The coupling of H-modes and ELMs is a typical example. One may not expect that a tokamak confinement could be pushed into the H-mode—an extreme status. Surprisingly, such an extreme confinement is not prone to total collapse, even if it is over-capacity. The ELMs limit the possible damage, making it edge localized and recoverable. This is like an amazing reservoir which one can fill to over its capacity, without worrying about its total collapse. The overflow damage is recoverable. This is the tokamak that we are working on. It may still reveal some unknown mystery features, bringing the hope of success.

Even though their role in stability has not been fully appreciated, the edge electric charges help in achieving H-mode confinement by stabilizing the instabilities. When the H-mode confinement slides toward collapse, the edge electric charges produce a quick ELM burst to release the overload locally, saving the system from total collapse. One may not expect the cooling of a tokamak plasma to result in heating, or that an Ohmic heating process may lead the plasma to spin or to rotate.

Fusion plasma theoretical research underwent a florescence in the 1980s or earlier. Many of the theories described in this book were developed in that period. Now, with construction of ITER and the rapid developments in tokamak control and diagnostic systems, the urgency to advance our understanding of the underlying physics has become more pressing than ever before.

doi:10.1088/978-1-6270-5423-2ch6

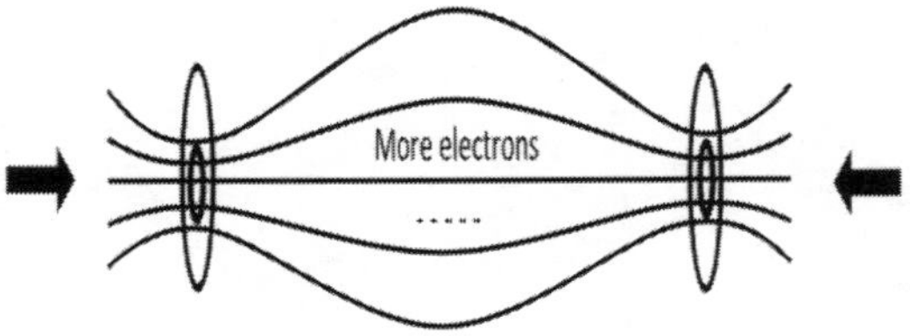

Figure 6.1. The partially non-neural mirror concept.

For a subject as difficult as controlled nuclear fusion, thinking smartly and thinking differently may help. ITER is often called 'the way'. Nonetheless, there may exist other possible ways. For example, the tokamak concept is not the only one that has the property of symmetry. The mirror concept can also be symmetric. Although the mirror has a loss cone, the introduction of extra electrons into the system, as shown in figure 6.1, can significantly reduce the loss cone. Because the electron mass is small, injecting the energetic electrons into the system and heating them perpendicularly does not require particularly high energy consumption. The excessive electrons can hold the ions in the loss cone both through negative potential and rotation, while the procession of the hot electrons can help in their own confinement. Particle procession can play a role as the tokamak safety factor to smooth the asymmetry defect. Also, multiple coil rings can be used to control the radial profile. Due to the plasma diamagnetic effect, ramping up the coil current can lead to plasma heating. Ohmic heating in tokamaks occurs through collisions. In the mirror case one can have direct heating. Ramping up the coil current at both ends can also pinch the mirror center plasma. A non-neutral ion mirror may be another possibility. In this case pinching and heating can both be applied to the ions. Readers are directed to [1] for more mirror features. Again, the uniqueness of the mirror concept lies in its beautiful symmetry and simple nature.

Reflecting on over half a century of efforts in this difficult field and the perhaps many years of effort ahead, I am reminded of Psalm 123:2.

As the eyes of slaves look to the hand of their master, as the eyes of a maid look to the hand of her mistress, so our eyes look to the LORD our God, till He shows us His mercy.

He may provide a clue...

Bibliography

[1] Post R 1987 The magnetic mirror approach to fusion *Nucl. Fusion* **27** 1579

Linjin Zheng

Appendix A

Derivation of the gyrokinetics equations

A.1 The first harmonic solution of the gyrokinetic equation

In this appendix, we detail the solutions of the first harmonic of the gyrokinetic equation (4.53) and compute their contributions to Ampere's law. The formal solution of the gyrokinetic equation has been given in (4.57). Here, we describe the explicit expressions. We denote the first harmonic contributions of individual terms on the right-hand side of (4.53) as $\delta\tilde{G}_{1a}$, $\delta\tilde{G}_{1b}$,..., with subscripts a, b,... representing the sequence order of the terms. The corresponding perturbed current moments are denoted by $\delta j_{a,\ldots;e_1,e_2} = \sum_{\mathrm{i,e}} e\int \mathrm{d}^3v(\mathbf{v}_\perp\cdot\mathbf{e}_{1,2})\delta\tilde{G}_{1a,\ldots}$.

The first harmonic contribution of the first term on the right-hand side of (4.54) is determined by

$$\mathcal{L}_g\delta\tilde{G}_{1a} = -\mathrm{i}\omega\frac{e}{m_\rho}\frac{\partial F_{g0}}{\partial\varepsilon}\mathbf{v}_\perp\cdot\delta\mathbf{A}(\mathbf{x}). \tag{A.1}$$

As will be seen, the MHD inertia effect is of order $\omega/\Omega \sim (\omega/\omega_{*\mathrm{i}})(\rho_\mathrm{i}^2/\lambda_\wedge L_\mathrm{p})$—the ratio of the first to second terms in (A.3) to be derived. Therefore, the solution of $\delta\tilde{G}_{1a}$ has to be carried out up to this order. Note that we have assumed $\omega \sim \omega_*$. The solution of (A.1) has to be carried out to the order which is order $\rho_\mathrm{i}^2/\lambda_\wedge L_\mathrm{p}$ smaller. This makes the calculations tedious.

Noting that μ, instead of $\mu_0 = v_\perp^2/2B$, is used as the adiabatic variable, equation (A.1) becomes

$$\begin{aligned}\mathcal{L}_g\delta\tilde{G}_{1a} = -\mathrm{i}\omega\frac{e}{m_\rho}\frac{\partial F_{g0}}{\partial\varepsilon}B^{3/2}\Bigg[&\sqrt{2\mu}\big(-\mathbf{e}_1(\mathbf{x})\sin\alpha + \mathbf{e}_2(x)\cos\alpha\big)\cdot\boldsymbol{\xi}(\mathbf{x})\\ &+\frac{1}{\sqrt{2\mu}}\mu_1\big(\mathbf{e}_1(\mathbf{x})\sin\alpha - \mathbf{e}_2(x)\cos\alpha\big)\cdot\boldsymbol{\xi}(\mathbf{x})\Bigg].\end{aligned} \tag{A.2}$$

doi:10.1088/978-1-6270-5423-2ch7

Let us first analyze the second term. This term is proportional to μ_1 and therefore is already of order ρ/L_B smaller. Inclusion of the next order correction to the perturbed quantity causes it to become of order $\rho^2/L_B\lambda_\perp$. Noting that $\rho^2/L_B\lambda_\perp \sim (\rho_i^2/\lambda_\wedge L_p)(\lambda_\wedge L_p/\lambda_\perp L_B)$, one can find that this term is of order $(\lambda_\wedge/\lambda_\perp)(L_p/L_B)$ compared to the inertia term, i.e. the second terms in (A.3) to be derived. Since this is not small, we keep their contributions and ignore further expansions. Likewise, in the solution of the first term on the right in (A.2) we keep the order $\rho^2/L_B\lambda_\perp$ contributions—as they are same order as the μ_1 contribution. This is justified for our intention of keeping only the second order FLR effects. The solution procedure of (A.2) is as follows: transforming the right-hand side to the guiding center coordinates, solving the resulting equation and transforming back the solution to the configuration coordinates. Noting that the $-\Omega\partial/\partial\alpha$ term is dominant, the rest of the terms in $\mathcal{L}_g$, for example $-\mathrm{i}\omega\delta\tilde{G}_{1a}$, can be solved perturbedly. The results are as follows

$$
\begin{aligned}
&\delta\tilde{G}_{1a}(\mathbf{x})\\
&= -\mathrm{i}\omega\frac{1}{B}\frac{\partial F_{g0}}{\partial\varepsilon}\mathbf{e}_b\times\mathbf{v}_\perp\cdot\delta\mathbf{A} + \frac{\omega^2}{\Omega}\frac{m_\rho}{T}F_{g0}v_\perp\big(-\sin\alpha\mathbf{e}_1+\cos\alpha\mathbf{e}_2\big)\cdot\boldsymbol{\xi}(\mathbf{x})\\
&\quad -\mathrm{i}\omega\frac{m_\rho}{T}F_{g0}v_\perp\cos\alpha\mathbf{e}_1\cdot\boldsymbol{\xi} - \mathrm{i}\omega\frac{m_\rho}{T}F_{g0}\frac{3v_\perp^3}{8\Omega^2}B^{-3/2}\cos\alpha\big\{\mathbf{e}_1\mathbf{e}_1+\mathbf{e}_2\mathbf{e}_2\big\}:\nabla\nabla\big(B^{3/2}\mathbf{e}_1\cdot\boldsymbol{\xi}\big)\\
&\quad -\mathrm{i}\omega\frac{m_\rho}{T}F_{g0}\cos\alpha\Bigg\{\mathbf{e}_1\cdot\nabla\mathbf{e}_1\cdot\boldsymbol{\xi}\Bigg[-\frac{5v_\perp}{8\Omega^2}\mathbf{e}_1\cdot\left(\frac{v_\perp^2}{2}\nabla\ln B+v_\parallel^2\boldsymbol{\kappa}\right)\\
&\quad +\frac{v_\perp^3}{8\Omega^2}\frac{\partial\ln n_0}{\partial\psi}\left[1+\eta\left(\frac{m_\rho\varepsilon}{T}-\frac{5}{2}\right)\right]-\frac{v_\perp^3}{16\Omega^2}\mathbf{e}_1\cdot(\nabla\ln B)\Bigg]\\
&\quad +\mathbf{e}_2\cdot\nabla\mathbf{e}_1\cdot\boldsymbol{\xi}\left[-\frac{5v_\perp}{8\Omega^2}\mathbf{e}_2\cdot\left(\frac{v_\perp^2}{2}\nabla\ln B+v_\parallel^2\boldsymbol{\kappa}\right)-\frac{9v_\perp^3}{16\Omega^2}\mathbf{e}_2\cdot(\nabla\ln B)\right]\\
&\quad +\frac{7v_\perp^3}{8\Omega^2}\left[\big(\mathbf{e}_1\cdot\nabla\mathbf{e}_1\big)\cdot\nabla\mathbf{e}_1\cdot\boldsymbol{\xi}+\big(\mathbf{e}_2\cdot\nabla\mathbf{e}_2\big)\cdot\nabla\mathbf{e}_1\cdot\boldsymbol{\xi}\right]\\
&\quad +\mathbf{e}_1\cdot\nabla\mathbf{e}_2\cdot\boldsymbol{\xi}\left[\frac{v_\perp}{8\Omega^2}\mathbf{e}_2\cdot\left(\frac{v_\perp^2}{2}\nabla\ln B+v_\parallel^2\boldsymbol{\kappa}\right)+\frac{7v_\perp^3}{16\Omega^2}\mathbf{e}_2\cdot(\nabla\ln B)\right]\\
&\quad +\mathbf{e}_2\cdot\nabla\mathbf{e}_2\cdot\boldsymbol{\xi}\Bigg[-\frac{v_\perp}{8\Omega^2}\mathbf{e}_1\cdot\left(\frac{v_\perp^2}{2}\nabla\ln B+v_\parallel^2\boldsymbol{\kappa}\right)-\frac{v_\perp^3}{8\Omega^2}\frac{\partial\ln n_0}{\partial\psi}\left[1+\eta\left(\frac{m_\rho\varepsilon}{T}-\frac{5}{2}\right)\right]\\
&\quad +\frac{v_\perp^3}{16\Omega^2}\mathbf{e}_1\cdot(\nabla\ln B)\Bigg]
\end{aligned}
$$

$$+\frac{v_\perp^3}{8\Omega^2}\Big[-(\mathbf{e}_2\cdot\nabla\mathbf{e}_1)\cdot\nabla\mathbf{e}_2\cdot\boldsymbol{\xi}+(\mathbf{e}_1\cdot\nabla\mathbf{e}_2)\cdot\nabla\mathbf{e}_2\cdot\boldsymbol{\xi}\Big]\Bigg\}$$

$$-\mathrm{i}\omega\frac{m_\rho}{T}F_{g0}v_\perp\sin\alpha\mathbf{e}_2\cdot\boldsymbol{\xi}-\mathrm{i}\omega\frac{m_\rho}{T}F_{g0}\frac{3v_\perp^3}{8\Omega^2}B^{-3/2}\sin\alpha\{\mathbf{e}_1\mathbf{e}_1+\mathbf{e}_2\mathbf{e}_2\}:\nabla\nabla\Big(B^{3/2}\mathbf{e}_2\cdot\boldsymbol{\xi}\Big)$$

$$-\mathrm{i}\omega\frac{m_\rho}{T}F_{g0}\sin\alpha\Bigg\{\mathbf{e}_1\cdot\nabla\mathbf{e}_1\cdot\boldsymbol{\xi}\Bigg[-\frac{v_\perp}{8\Omega^2}\mathbf{e}_2\cdot\Bigg(\frac{v_\perp^2}{2}\nabla\ln B+v_\parallel^2\boldsymbol{\kappa}\Bigg)+\frac{v_\perp^3}{16\Omega^2}\mathbf{e}_2\cdot(\nabla\ln B)\Bigg]$$

$$+\mathbf{e}_2\cdot\nabla\mathbf{e}_1\cdot\boldsymbol{\xi}\Bigg[\frac{v_\perp}{8\Omega^2}\mathbf{e}_1\cdot\Bigg(\frac{v_\perp^2}{2}\nabla\ln B+v_\parallel^2\boldsymbol{\kappa}\Bigg)-\frac{3v_\perp^3}{8\Omega^2}\frac{\partial\ln n_0}{\partial\psi}\Bigg[1+\eta\Bigg(\frac{m_\rho\varepsilon}{T}-\frac{5}{2}\Bigg)\Bigg]$$

$$+\frac{7v_\perp^3}{16\Omega^2}\mathbf{e}_1\cdot(\nabla\ln B)\Bigg]$$

$$+\frac{v_\perp^3}{8\Omega^2}\Big[(\mathbf{e}_2\cdot\nabla\mathbf{e}_1)\cdot\nabla\mathbf{e}_1\cdot\boldsymbol{\xi}-(\mathbf{e}_1\cdot\nabla\mathbf{e}_2)\cdot\nabla\mathbf{e}_1\cdot\boldsymbol{\xi}\Big]$$

$$+\mathbf{e}_1\cdot\nabla\mathbf{e}_2\cdot\boldsymbol{\xi}\Bigg[-\frac{5v_\perp}{8\Omega^2}\mathbf{e}_1\cdot\Bigg(\frac{v_\perp^2}{2}\nabla\ln B+v_\parallel^2\boldsymbol{\kappa}\Bigg)+\frac{5v_\perp^3}{8\Omega^2}\frac{\partial\ln n_0}{\partial\psi}\Bigg[1+\eta\Bigg(\frac{m_\rho\varepsilon}{T}-\frac{5}{2}\Bigg)\Bigg]$$

$$-\frac{9v_\perp^3}{16\Omega^2}\mathbf{e}_1\cdot(\nabla\ln B)\Bigg]$$

$$+\mathbf{e}_2\cdot\nabla\mathbf{e}_2\cdot\boldsymbol{\xi}\Bigg[-\frac{5v_\perp}{8\Omega^2}\mathbf{e}_2\cdot\Bigg(\frac{v_\perp^2}{2}\nabla\ln B+v_\parallel^2\boldsymbol{\kappa}\Bigg)-\frac{v_\perp^3}{16\Omega^2}\mathbf{e}_2\cdot(\nabla\ln B)\Bigg]$$

$$+\frac{7v_\perp^3}{8\Omega^2}\Big[(\mathbf{e}_1\cdot\nabla\mathbf{e}_1)\cdot\nabla\mathbf{e}_2\cdot\boldsymbol{\xi}+(\mathbf{e}_2\cdot\nabla\mathbf{e}_2)\cdot\nabla\mathbf{e}_2\cdot\boldsymbol{\xi}\Big]\Bigg\}. \tag{A.3}$$

Here, the second term on the right-hand side derives from the first order correction from $-\mathrm{i}\omega\delta\tilde{G}_{1a}^{(0)}$, where $\delta\tilde{G}_{1a}^{(0)}$ represents the first term on the right-hand side of (A.3). This term, as will be seen, gives rise to the MHD inertia term. In obtaining (A.3), the contribution from the term $\dot{\alpha}_1(\partial\delta\tilde{G}_{1a}^{(0)}/\partial\alpha)$ is also retained, but the first harmonic contribution of the term $\dot{\mathbf{X}}\cdot\nabla_X\delta\tilde{G}_{1a}$ is dropped, as it is odd in $v_\parallel$.

Using the quasi-neutrality condition, one can prove that the leading order contribution of $\delta\tilde{G}_{1a}$ (i.e. the first term on the right-hand side of (A.3)) to the current moment vanishes. The current moments from $\delta\tilde{G}_{1a}$ are then given as follows

$$
\begin{aligned}
&\delta j_{1a;\mathbf{e}_1}\\
&=\frac{\omega^2 n_0 m_\rho}{B}\mathbf{e}_2\cdot\boldsymbol{\xi}(\mathbf{x})\\
&\quad-\mathrm{i}\omega\frac{2n_0 Te_\mathrm{i}}{m_\rho}\frac{3}{4\Omega_\mathrm{i}^2}B^{-3/2}\{\mathbf{e}_1\mathbf{e}_1+\mathbf{e}_2\mathbf{e}_2\}:\nabla\nabla\left(B^{3/2}\mathbf{e}_1\cdot\boldsymbol{\xi}\right)\\
&\quad-\mathrm{i}\omega\frac{2n_0 Te_\mathrm{i}}{m_\rho}\left\{\mathbf{e}_1\cdot\nabla\mathbf{e}_1\cdot\boldsymbol{\xi}\left[-\frac{5}{8\Omega_\mathrm{i}^2}\mathbf{e}_1\cdot\left(\nabla\ln B+\frac{1}{2}\boldsymbol{\kappa}\right)+\frac{|\nabla\psi|}{4\Omega_\mathrm{i}^2}\frac{\partial\ln n_0}{\partial\psi}\left(1+\eta_\mathrm{i}\right)\right.\right.\\
&\quad\left.-\frac{1}{8\Omega_\mathrm{i}^2}\mathbf{e}_1\cdot(\nabla\ln B)\right]\\
&\quad+\mathbf{e}_2\cdot\nabla\mathbf{e}_1\cdot\boldsymbol{\xi}\left[-\frac{5}{8\Omega_\mathrm{i}^2}\mathbf{e}_2\cdot\left(\nabla\ln B+\frac{1}{2}\boldsymbol{\kappa}\right)-\frac{9}{8\Omega_\mathrm{i}^2}\mathbf{e}_2\cdot(\nabla\ln B)\right]\\
&\quad+\frac{7}{4\Omega_\mathrm{i}^2}\left[\left(\mathbf{e}_1\cdot\nabla\mathbf{e}_1\right)\cdot\nabla\mathbf{e}_1\cdot\boldsymbol{\xi}+\left(\mathbf{e}_2\cdot\nabla\mathbf{e}_2\right)\cdot\nabla\mathbf{e}_1\cdot\boldsymbol{\xi}\right]\\
&\quad+\mathbf{e}_1\cdot\nabla\mathbf{e}_2\cdot\boldsymbol{\xi}\left[\frac{1}{8\Omega_\mathrm{i}^2}\mathbf{e}_2\cdot\left(\nabla\ln B+\frac{1}{2}\boldsymbol{\kappa}\right)+\frac{7}{8\Omega_\mathrm{i}^2}\mathbf{e}_2\cdot(\nabla\ln B)\right]\\
&\quad+\mathbf{e}_2\cdot\nabla\mathbf{e}_2\cdot\boldsymbol{\xi}\left[-\frac{1}{8\Omega_\mathrm{i}^2}\mathbf{e}_1\cdot\left(\nabla\ln B+\frac{1}{2}\boldsymbol{\kappa}\right)-\frac{|\nabla\psi|}{4\Omega_\mathrm{i}^2}\frac{\partial\ln n_0}{\partial\psi}\left(1+\eta_\mathrm{i}\right)\right.\\
&\quad\left.\left.+\frac{1}{8\Omega_\mathrm{i}^2}\mathbf{e}_1\cdot(\nabla\ln B)\right]+\frac{1}{4\Omega_\mathrm{i}^2}\left[-\left(\mathbf{e}_2\cdot\nabla\mathbf{e}_1\right)\cdot\nabla\mathbf{e}_2\cdot\boldsymbol{\xi}+\left(\mathbf{e}_1\cdot\nabla\mathbf{e}_2\right)\cdot\nabla\mathbf{e}_2\cdot\boldsymbol{\xi}\right]\right\},
\end{aligned}
\tag{A.4}
$$

$$
\begin{aligned}
&\delta j_{1a;\mathbf{e}_2}\\
&=-\frac{\omega^2 n_0 m_\rho}{B}\mathbf{e}_1\cdot\boldsymbol{\xi}\\
&\quad-\mathrm{i}\omega\frac{2n_0 Te_\mathrm{i}}{m_\rho}\frac{3}{4\Omega_\mathrm{i}^2}B^{-3/2}\{\mathbf{e}_1\mathbf{e}_1+\mathbf{e}_2\mathbf{e}_2\}:\nabla\nabla\left(B^{3/2}\mathbf{e}_2\cdot\boldsymbol{\xi}\right)\\
&\quad-\mathrm{i}\omega\frac{2n_0 T_\mathrm{i} e_\mathrm{i}}{m_{\rho\mathrm{i}}}\left\{\mathbf{e}_1\cdot\nabla\mathbf{e}_1\cdot\boldsymbol{\xi}\left[-\frac{1}{8\Omega_\mathrm{i}^2}\mathbf{e}_2\cdot\left(\nabla\ln B+\frac{1}{2}\boldsymbol{\kappa}\right)+\frac{1}{8\Omega_\mathrm{i}^2}\mathbf{e}_2\cdot(\nabla\ln B)\right]\right.
\end{aligned}
$$

$$+ \mathbf{e}_2 \cdot \nabla \mathbf{e}_1 \cdot \boldsymbol{\xi} \left[\frac{1}{8\Omega_{\rm i}^2} \mathbf{e}_1 \cdot \left(\nabla \ln B + \frac{1}{2} \boldsymbol{\kappa} \right) - \frac{3|\nabla \psi|}{4\Omega_{\rm i}^2} \frac{\partial \ln n_0}{\partial \psi} \left(1 + \eta_{\rm i} \right) \right.$$

$$\left. + \frac{7}{8\Omega_{\rm i}^2} \mathbf{e}_1 \cdot (\nabla \ln B) \right]$$

$$+ \frac{1}{4\Omega_{\rm i}^2} \left[\left(\mathbf{e}_2 \cdot \nabla \mathbf{e}_1 \right) \cdot \nabla \mathbf{e}_1 \cdot \boldsymbol{\xi} - \left(\mathbf{e}_1 \cdot \nabla \mathbf{e}_2 \right) \cdot \nabla \mathbf{e}_1 \cdot \boldsymbol{\xi} \right]$$

$$+ \mathbf{e}_1 \cdot \nabla \mathbf{e}_2 \cdot \boldsymbol{\xi} \left[-\frac{5}{8\Omega_{\rm i}^2} \mathbf{e}_1 \cdot \left(\nabla \ln B + \frac{1}{2} \boldsymbol{\kappa} \right) + \frac{5|\nabla \psi|}{4\Omega_{\rm i}^2} \frac{\partial \ln n_0}{\partial \psi} \left(1 + \eta_{\rm i} \right) \right.$$

$$\left. - \frac{9}{8\Omega_{\rm i}^2} \mathbf{e}_1 \cdot (\nabla \ln B) \right]$$

$$+ \mathbf{e}_2 \cdot \nabla \mathbf{e}_2 \cdot \boldsymbol{\xi} \left[-\frac{5}{8\Omega_{\rm i}^2} \mathbf{e}_2 \cdot \left(\nabla \ln B + \frac{1}{2} \boldsymbol{\kappa} \right) - \frac{1}{8\Omega_{\rm i}^2} \mathbf{e}_2 \cdot (\nabla \ln B) \right]$$

$$\left. + \frac{7}{4\Omega_{\rm i}^2} \left[\left(\mathbf{e}_1 \cdot \nabla \mathbf{e}_1 \right) \cdot \nabla \mathbf{e}_2 \cdot \boldsymbol{\xi} + \left(\mathbf{e}_2 \cdot \nabla \mathbf{e}_2 \right) \cdot \nabla \mathbf{e}_2 \cdot \boldsymbol{\xi} \right] \right\}. \tag{A.5}$$

Here, one can see that the first terms on the right-hand side of (A.4) and (A.5) correspond to the ideal MHD inertia term. The second term on the right is formally of order $(\omega_{*\rm i}/\omega)(L_{\rm p}/\lambda_{\perp})$ larger than the ideal MHD inertia term. But, in the vorticity equation $\nabla \cdot \delta \mathbf{j} = 0$ to be derived in appendix A.3, they become of the same order for the case with a mode frequency much lower than the compressional Alfvén mode frequency. In the case with a mode frequency of the same order as the compressional Alfvén mode frequency, one has $\omega \gg \omega_{*\rm i}$ and the second term becomes negligible. The rest terms are of the same order as the inertia term.

The first harmonic contribution of the second term on the right-hand side of (4.54), $\delta \tilde{G}_{1b}$, is of order $(L_{\rm p}/L_B)(\lambda_{\wedge}/\rho)(\omega_{*\rm i}/\omega)^2$ compared to the second term on the right-hand side of (A.3). But, its leading order is odd in $v_{\parallel}$ and therefore it makes no contribution to the perturbed current moment. The next non-vanishing correction of the first harmonic is reduced by order $\rho^2/\lambda_{\perp}^2$. Therefore, the first harmonic contribution of the second term on the right-hand side of (4.54) can be ignored compared to the inertia term.

The first harmonic contribution of the third term on the right-hand side of (4.54) is determined by

$$-\Omega \frac{\partial \delta \tilde{G}_{1c}}{\partial \alpha} = -{\rm i}\omega |\nabla \psi| \frac{\partial F_{g0}}{\partial \psi} \frac{1}{\Omega} \mathbf{v} \times \mathbf{e}_b \cdot \nabla \left(\mathbf{e}_1 \cdot \boldsymbol{\xi} \right).$$

One can find that $\delta\tilde{G}_{1c}$ is of order ω_{*i}/ω compared to the second term on the right-hand side of (A.3). Therefore, only the lowest order solution needs to be determined

$$\delta\tilde{G}_{1c} = -i\omega\frac{|\nabla\psi|}{\Omega^2}\frac{\partial F_{g0}}{\partial\psi}\mathbf{v}_\perp \cdot \nabla\left(\mathbf{e}_1 \cdot \boldsymbol{\xi}\right).$$

The corresponding current moment is given by

$$\delta j_{1c;e_1,e_2} = -i\omega\frac{|\nabla\psi|}{B\Omega_i}P_i'\mathbf{e}_{1,2} \cdot \nabla\left(\mathbf{e}_1 \cdot \boldsymbol{\xi}\right).$$

This term gives rise to the so-called $\omega_{*i}(1+\eta_i)$ modification to the ideal MHD inertia term, noting that $\mathbf{e}_1 \cdot \nabla(\mathbf{e}_1 \cdot \boldsymbol{\xi}) \approx -\mathbf{e}_2 \cdot \nabla(\mathbf{e}_2 \cdot \boldsymbol{\xi})$ for the modes with frequencies much lower than the compressional Alfvén frequency.

The first harmonic contribution of the forth term on the right-hand side of (4.54) is determined by

$$-\Omega\frac{\partial\delta\tilde{G}_{1d}}{\partial\alpha} = -i\left(\omega - \omega_*^T\right)\frac{e}{m_\rho}\frac{\partial F_{g0}}{\partial\varepsilon}\frac{1}{\Omega}\mathbf{v} \times \mathbf{e}_b \cdot \nabla\delta\varphi(\mathbf{X}),$$

where the guiding center expansion of $\delta\varphi(\mathbf{x})$ has been made and only the leading order contribution is kept. The solution is

$$\delta\tilde{G}_{1d} = -i\frac{1}{B\Omega}\left(\omega - \omega_*^T\right)\frac{\partial F_{g0}}{\partial\varepsilon}\mathbf{v}_\perp \cdot \nabla\delta\varphi(\mathbf{x}). \tag{A.6}$$

The leading order contribution of $\delta\tilde{G}_{1d}$ to the two perpendicular components of the perturbed current moment are as follows

$$\delta j_{1d;e_1,e_2} = in_0 m_i\left[\omega - \omega_{*i}\left(1 + \eta_i\right)\right]\frac{1}{B^2}\mathbf{e}_{1,2} \cdot \nabla\delta\varphi.$$

This term represents the parallel-electric-field modification of the MHD inertia term. This term can be of the same order as the inertia term near the singular layers.

The first harmonic contribution of the fifth term on the right-hand side of (4.54) is of order $\lambda_\wedge/L_B$ compared to $\delta\tilde{G}_{1d}$ in (A.6) and therefore can be neglected.

The first harmonic contribution of the sixth term on the right-hand side of (4.54) is determined by

$$-\Omega\frac{\partial\delta\tilde{G}_{1f}}{\partial\alpha} = \frac{e}{m_\rho}\left(\frac{g}{B}\frac{\partial F_{g0}}{\partial\Psi} + \left|v_\parallel\right| B\frac{\partial\bar{F}_{g1}}{\partial\mu}\right)\mathbf{e}_b \times \mathbf{v}_\perp \cdot \delta\mathbf{B}(\mathbf{x}). \tag{A.7}$$

First, we note that, according to (4.15), the two terms in parenthesis are of the same order. One can find that $\delta\tilde{G}_{1f}$ is of order $(\lambda_\wedge^2/\rho_i^2)(\omega_{*i}/\omega)^2$ compared to the second term on the right-hand side of (A.3). Here, an order estimate, $\delta\mathbf{B}_\perp \sim (B/L_B)\xi_\wedge$, has been used. Therefore, the solution of (A.7) has to be carried out up to the second order, yielding

$$\delta\tilde{G}_{1f}(\mathbf{x}) = -\frac{1}{\Omega}\left(\frac{g}{B}\frac{\partial F_{g0}}{\partial\psi} + \left|v_{\parallel}\right|\frac{e}{m_\rho B}\frac{\partial\bar{F}_{g1}}{\partial\mu}\right)\mathbf{v}_\perp\cdot\delta\mathbf{B}(\mathbf{x})$$
$$-\frac{3}{8\Omega^3}\left(\frac{g}{B}\frac{\partial F_{g0}}{\partial\psi} + \left|v_{\parallel}\right|\frac{e}{m_\rho B}\frac{\partial\bar{F}_{g1}}{\partial\mu}\right)v_\perp^2(\mathbf{e}_1\mathbf{e}_1 + \mathbf{e}_2\mathbf{e}_2) : \nabla\nabla\left(\mathbf{v}_\perp\cdot\delta\mathbf{B}(\mathbf{x})\right).$$

The current moment is given by

$$\delta j_{1f;e_1,e_2} = -\frac{gP'}{B^2}\mathbf{e}_{1,2}\cdot\delta\mathbf{B} - g'\mathbf{e}_{1,2}\cdot\delta\mathbf{B}$$
$$-\frac{3}{2\Omega_\mathrm{i}}\left(\frac{1}{B^2}\frac{gn_0T_\mathrm{i}^2}{\Omega_\mathrm{i}m_{\rho\mathrm{i}}}\frac{\partial\ln n_0}{\partial\psi}\left(1+2\eta_\mathrm{i}\right) - 4\pi m_{\rho\mathrm{i}}\int \mathrm{d}\varepsilon\mathrm{d}\mu_0\mu_0\bar{F}_{g1}\right)$$
$$\times(\mathbf{e}_1\mathbf{e}_1 + \mathbf{e}_2\mathbf{e}_2) : \nabla\nabla\left(\mathbf{e}_{1,2}\cdot\delta\mathbf{B}\right).$$

The first two terms on the right are the ideal MHD effects. This shows the necessity of keeping the neoclassical correction of equilibrium distribution function, even for recovering the ideal MHD.

The first harmonic contribution of the seventh term on the right-hand side of (4.54) is determined by

$$-\Omega\frac{\partial\delta\tilde{G}_{1g}}{\partial\alpha} = \frac{1}{B}\mathbf{v}_\perp\cdot\nabla_X F_{g0}\mathbf{e}_b\cdot\delta\mathbf{B}(\mathbf{x}). \tag{A.8}$$

One can find that $\delta\tilde{G}_{1g}$ is of order $(P/B^2)(\lambda_\wedge\lambda_\perp/\rho_\mathrm{i}^2)(\omega_{*\mathrm{i}}/\omega)^2$ compared to the second term on the right-hand side of (A.3). Here, the order estimate: $\delta\mathbf{B}_\parallel \sim B\,\nabla\cdot\boldsymbol{\xi}_\perp \sim (B/L_B)\xi_\wedge$ is employed, which corresponds to the case with a mode frequency much lower than the compressional Alfvén mode frequency. In the case with a mode frequency of the same order as ω_*, the solution of (A.8) has to be carried out up to the second order, yielding

$$\delta\tilde{G}_{1g}(\mathbf{x}) = -\frac{|\nabla\psi|}{B\Omega}\frac{\partial F_{g0}}{\partial\psi}v_\perp\sin\alpha\mathbf{e}_b\cdot\delta\mathbf{B}(\mathbf{x})$$
$$-\frac{3|\nabla\psi|}{8\Omega^3 B}\frac{\partial F_{g0}}{\partial\psi}v_\perp^3\sin\alpha(\mathbf{e}_1\mathbf{e}_1 + \mathbf{e}_2\mathbf{e}_2) : \nabla\nabla\left(\mathbf{e}_b\cdot\delta\mathbf{B}\right).$$

The current moment is given by

$$\delta j_{1g;e_1} = 0,$$

$$\delta j_{1g;e_2} = -\frac{P'|\nabla\psi|}{B^2}\delta\mathbf{B}\cdot\mathbf{e}_b - \frac{3|\nabla\psi|\,e_\mathrm{i}n_0}{2B\Omega_\mathrm{i}^3}\left(\frac{T_\mathrm{i}}{m_{\rho\mathrm{i}}}\right)^2\frac{\partial\ln n_0}{\partial\psi}\left(1+2\eta_\mathrm{i}\right)$$
$$\times(\mathbf{e}_1\mathbf{e}_1 + \mathbf{e}_2\mathbf{e}_2) : \nabla\nabla\left(\mathbf{e}_b\cdot\delta\mathbf{B}\right).$$

The eighth term on the right-hand side of (4.54) is of order $(\omega_{*i}/\omega)(\lambda_\wedge/L_p)$ compared to the forth term on the right-hand side of (4.54). Therefore, $\delta\tilde{G}_{1h}$ can be dropped. Although in developing the interchange mode theories it is allowable to consider $\lambda_\wedge \sim L_p$, we do not intend to include the FLR corrections to the modes with long geodesic wavelength $\lambda_\wedge$. The diamagnetic frequency is inversely proportional to $\lambda_\wedge$ and therefore is small in this case.

The ninth term on the right-hand side of (4.54) is of the same order as the term giving rise to the second term on the right-hand side of (A.3). But, its leading order is odd in $v_\parallel$ and therefore makes no contribution to the perturbed current moment.

The combination of the tenth and eleventh terms is of order ω_{si}/ω. But, it is odd in $v_\parallel$ and therefore makes no contribution to the perturbed current moment.

In addition, we also need to consider the first harmonic contribution of the convective term $-\boldsymbol{\xi}(\mathbf{X}) \cdot \nabla F_{g0}$—i.e. the second term on the right-hand side of (4.52). Note that $\boldsymbol{\xi}(\mathbf{X})$ in the convective term is given in the guiding center coordinates. Converting back to the particle coordinates, the first harmonic contribution of the convective term becomes

$$\begin{aligned}
\delta\tilde{G}_{1\text{conv}} = & -\frac{v_\perp}{\Omega}\big(\mathbf{e}_1 \sin\alpha - \mathbf{e}_2\cos\alpha\big) \cdot \nabla\boldsymbol{\xi} \cdot \nabla F_{g0} \\
& - \frac{v_\perp^3}{8\Omega^3}\,|\nabla\psi|\,\frac{\partial F_{g0}}{\partial\psi}\big(\mathbf{e}_1\mathbf{e}_1\mathbf{e}_1 \sin\alpha + \mathbf{e}_1\mathbf{e}_2\mathbf{e}_2\sin\alpha - \mathbf{e}_1\mathbf{e}_1\mathbf{e}_2\cos\alpha - \mathbf{e}_2\mathbf{e}_2\mathbf{e}_2\cos\alpha\big) \\
& \vdots\, \nabla\nabla\nabla\big(\boldsymbol{\xi}\cdot\mathbf{e}_1\big) \\
& - \frac{v_\perp^3}{8\Omega^3}\frac{\partial F_{g0}}{\partial\psi}\mathbf{e}_1 \cdot \nabla(|\nabla\psi|)\big(3\mathbf{e}_1\mathbf{e}_1\sin\alpha + \mathbf{e}_2\mathbf{e}_2\sin\alpha - 2\mathbf{e}_1\mathbf{e}_2\cos\alpha\big) : \nabla\nabla\big(\boldsymbol{\xi}\cdot\mathbf{e}_1\big) \\
& - \frac{v_\perp^3}{8\Omega^3}\frac{\partial F_{g0}}{\partial\psi}\mathbf{e}_2 \cdot \nabla(|\nabla\psi|)\big(2\mathbf{e}_1\mathbf{e}_2\sin\alpha - \mathbf{e}_1\mathbf{e}_1\cos\alpha - 3\mathbf{e}_2\mathbf{e}_2\cos\alpha\big) : \nabla\nabla\big(\boldsymbol{\xi}\cdot\mathbf{e}_1\big) \\
& - \frac{v_\perp^3}{8\Omega^3}|\nabla\psi|\,\mathbf{e}_1\cdot\nabla\left(\frac{\partial F_{g0}}{\partial\psi}\right)\big(3\mathbf{e}_1\mathbf{e}_1\sin\alpha + \mathbf{e}_2\mathbf{e}_2\sin\alpha - 2\mathbf{e}_1\mathbf{e}_2\cos\alpha\big) : \nabla\nabla\big(\boldsymbol{\xi}\cdot\mathbf{e}_1\big).
\end{aligned}$$

Here, we note that the first term of this solution is of order $(L_p\lambda_\wedge/\rho_i^2)(\omega_{*i}/\omega)^2$ compared to the inertia term, the second term on the right-hand side of (A.3), and the second term is a correction of order $(\rho_i/\lambda_\perp)^2$ to the first term. The two components of the $\delta\tilde{G}_{1\text{conv}}$-induced current moment are given as follows

$$\begin{aligned}
\delta j_{1\text{conv};\mathbf{e}_1} = & \frac{1}{B}\mathbf{e}_2 \cdot \nabla\boldsymbol{\xi} \cdot \nabla P \\
& + \frac{n_0 e_i}{2\Omega_i^3}\left(\frac{T}{m_{\rho i}}\right)^2 |\nabla\psi|\,\frac{\partial \ln n_0}{\partial\psi}(1 + 2\eta)(\mathbf{e}_1\mathbf{e}_1\mathbf{e}_2 + \mathbf{e}_2\mathbf{e}_2\mathbf{e}_2)\,\vdots\,\nabla\nabla\nabla\big(\boldsymbol{\xi}\cdot\mathbf{e}_1\big)
\end{aligned}$$

$$+\frac{n_0 e_{\rm i}}{\Omega_{\rm i}^3}\left(\frac{T}{m_{\rho {\rm i}}}\right)^2\frac{\partial \ln n_0}{\partial\psi}(1+2\eta)\mathbf{e}_1\cdot\nabla(|\nabla\psi|)\mathbf{e}_1\mathbf{e}_2:\nabla\nabla\big(\boldsymbol{\xi}\cdot\mathbf{e}_1\big)$$

$$+\frac{n_0 e_{\rm i}}{2\Omega_{\rm i}^3}\left(\frac{T}{m_{\rho {\rm i}}}\right)^2\frac{\partial \ln n_0}{\partial\psi}(1+2\eta)\mathbf{e}_2\cdot\nabla(|\nabla\psi|)\big(\mathbf{e}_1\mathbf{e}_1+3\mathbf{e}_2\mathbf{e}_2\big):\nabla\nabla\big(\boldsymbol{\xi}\cdot\mathbf{e}_1\big)$$

$$+\frac{n_0 e_{\rm i}}{\Omega_{\rm i}^3}\left(\frac{T}{m_{\rho {\rm i}}}\right)^2|\nabla\psi|^2\left\{\left(\frac{\partial \ln n_0}{\partial\psi}\right)^2\left(1+\frac{15}{2}\eta^2+4\eta\right)\right.$$

$$\left.+\frac{\partial^2 \ln n_0}{\partial\psi^2}(1+2\eta)+\frac{\partial \ln n_0}{\partial\psi}\left(2\frac{\partial\eta}{\partial\psi}-\eta\frac{7}{2}\frac{\partial \ln T}{\partial\psi}\right)\right\}\mathbf{e}_1\mathbf{e}_2:\nabla\nabla\big(\boldsymbol{\xi}\cdot\mathbf{e}_1\big),$$

(A.9)

$$\delta j_{1{\rm conv};\mathbf{e}_2}=-\frac{1}{B}\mathbf{e}_1\cdot\nabla\boldsymbol{\xi}\cdot\nabla P$$

$$-\frac{n_0 e_{\rm i}}{2\Omega_{\rm i}^3}\left(\frac{T}{m_{\rho {\rm i}}}\right)^2|\nabla\psi|\frac{\partial \ln n_0}{\partial\psi}(1+2\eta)(\mathbf{e}_1\mathbf{e}_1\mathbf{e}_1+\mathbf{e}_1\mathbf{e}_2\mathbf{e}_2):\nabla\nabla\nabla\big(\boldsymbol{\xi}\cdot\mathbf{e}_1\big)$$

$$-\frac{n_0 e_{\rm i}}{2\Omega_{\rm i}^3}\left(\frac{T}{m_{\rho {\rm i}}}\right)^2\frac{\partial \ln n_0}{\partial\psi}(1+2\eta)\mathbf{e}_1\cdot\nabla(|\nabla\psi|)\big(3\mathbf{e}_1\mathbf{e}_1+\mathbf{e}_2\mathbf{e}_2\big):\nabla\nabla\big(\boldsymbol{\xi}\cdot\mathbf{e}_1\big)$$

$$-\frac{n_0 e_{\rm i}}{\Omega_{\rm i}^3}\left(\frac{T}{m_{\rho {\rm i}}}\right)^2\frac{\partial \ln n_0}{\partial\psi}(1+2\eta)\mathbf{e}_2\cdot\nabla(|\nabla\psi|)\mathbf{e}_1\mathbf{e}_2:\nabla\nabla\big(\boldsymbol{\xi}\cdot\mathbf{e}_1\big)$$

$$-\frac{n_0 e_{\rm i}}{2\Omega_{\rm i}^3}\left(\frac{T}{m_{\rho {\rm i}}}\right)^2|\nabla\psi|^2\left\{\left(\frac{\partial \ln n_0}{\partial\psi}\right)^2\left(1+\frac{15}{2}\eta^2+4\eta\right)\right.$$

$$\left.+\frac{\partial^2 \ln n_0}{\partial\psi^2}(1+2\eta)+\frac{\partial \ln n_0}{\partial\psi}\left(2\frac{\partial\eta}{\partial\psi}-\eta\frac{7}{2}\frac{\partial \ln T}{\partial\psi}\right)\right\}$$

$$\times(\mathbf{e}_1\mathbf{e}_1+\mathbf{e}_2\mathbf{e}_2):\nabla\nabla\big(\boldsymbol{\xi}\cdot\mathbf{e}_1\big).\qquad\text{(A.10)}$$

Here, the second terms on the right-hand sides of (A.9) and (A.10) are formally of order $(\omega_{*{\rm i}}/\omega)^2(L_{\rm p}/\lambda)$ larger than the ideal MHD inertia term. But, in the vorticity equation $\nabla\cdot\delta\mathbf{j}=0$ to be derived in appendix A.3, it becomes of the same order as the inertia term, noting that the $(\mathbf{e}_1\cdot\nabla\mathbf{e}_2\cdot\nabla-\mathbf{e}_2\cdot\nabla\mathbf{e}_1\cdot\nabla)\mathbf{e}_1\cdot\boldsymbol{\xi}(\mathbf{x})$ cancels itself in the leading order.

Collecting the results in this appendix, one obtains the first harmonic solution of the gyrokinetic equation in (4.60), and Ampere's law in (4.63) and (4.64).

A.2 The gyrophase-averaged gyrokinetic equation

In this appendix, we show the reduction of the gyrophase-averaged gyrokinetic equation, (4.61), for calculating the pressure moment in Ampere's law and the density moment in the quasi-neutrality condition.

First, we note that only the even part $[\delta G_0^{\mathrm{e}} = (\delta G_0(v_\parallel) + \delta G_0(-v_\parallel))/2]$ of the distribution function with respect to the parallel velocity is necessary for calculating the pressure and density moments. Inspecting the right-hand side of (4.53), one can see that $\mathcal{R}$ in (4.54) contains both even ($\mathcal{R}^{\mathrm{e}}$) and odd ($\mathcal{R}^{\mathrm{o}}$) parts with respect to $v_\parallel$. The gyrophase-averaged gyrokinetic equation for the even part can be formed as follows

$$v_\parallel \mathbf{e}_b \cdot \nabla \frac{v_\parallel}{\mathrm{i}(\omega - \omega_{\mathrm{d}})} \mathbf{e}_b \cdot \nabla \delta G_0^{\mathrm{e}} - \mathrm{i}(\omega - \omega_{\mathrm{d}})\delta G_0^{\mathrm{e}} = \mathcal{R}^{\mathrm{e}} + v_\parallel \mathbf{e}_b \cdot \nabla \frac{1}{\mathrm{i}(\omega - \omega_{\mathrm{d}})} \mathcal{R}^{\mathrm{o}}. \tag{A.11}$$

This expression shows that the contribution of the even part ($\mathcal{R}^{\mathrm{e}}$) is of order $(\omega/\omega_{\mathrm{t}})\mathcal{R}^{\mathrm{e}}$. Therefore, the odd part ($\mathcal{R}^{\mathrm{o}}$) is enhanced by order $(\lambda_\wedge/\rho)(L_{\mathrm{p}}/L_B)(\omega_*/\omega)$ compared to the even part ($\mathcal{R}^{\mathrm{e}}$). Here, ω_{t} represents the transit frequency. With these ordering estimates, we can analyze (4.54) term by term.

The gyrophase-average $\langle\cdots\rangle_\alpha$ of the first term on the right-hand side of (4.54) yields

$$\left\langle -\mathrm{i}\omega \frac{e}{m_\rho} \frac{\partial F_{g0}}{\partial \varepsilon} \mathbf{v}_\perp \cdot \delta \mathbf{A}(\mathbf{x}) \right\rangle_\alpha$$
$$= -\mathrm{i}\omega \mu_0 B \frac{\partial F_{g0}}{\partial \varepsilon} \nabla_\perp \cdot \boldsymbol{\xi} - \mathrm{i}\omega \mu_0 \frac{\partial F_{g0}}{\partial \varepsilon} \boldsymbol{\xi} \cdot \nabla B - \mathrm{i}\omega \frac{\partial F_{g0}}{\partial \varepsilon} \left(\mu_0 B - \frac{v_\parallel^2}{2} \right) \boldsymbol{\kappa} \cdot \boldsymbol{\xi}. \tag{A.12}$$

(Here, $\mu_0 = v_\perp^2/2B^2$, not to be confused with the magnetic constant.) Note that, in the case with the compressional Alfvén mode suppressed, one has $\nabla_\perp \cdot \boldsymbol{\xi} \sim \xi_\wedge/L_B$. Therefore, the calculation has been carried out one order further in (A.12). As discussed in the main text, the first term on the right-hand side corresponds to the term on the right-hand side of (2.8), which gives rise to the so-called apparent mass effect. Therefore, the first term (with $\nabla_\perp \cdot \boldsymbol{\xi} \sim \xi_\wedge/L_B$ assumed) can be used as a reference term.

There is one more contribution of the same type as (A.12). Note that $\delta \tilde{G}_{1a}$ in (A.3) can couple to the $\dot{\alpha}_1(\partial/\partial\alpha)$ term, yielding

$$-\left\langle \left(\mathbf{v} \cdot \nabla_x \alpha + \frac{1}{\Omega} \mathbf{v} \times \mathbf{e}_b \cdot \nabla_x \Omega \right) \frac{\partial \delta \tilde{G}_{1a}}{\partial \alpha} \right\rangle_\alpha = \mathrm{i}\omega \mu_0 \frac{\partial F_{g0}}{\partial \varepsilon} \boldsymbol{\xi} \cdot \nabla B + \mathrm{i}\omega \frac{\partial F_{g0}}{\partial \varepsilon} \frac{v_\parallel^2}{2} \boldsymbol{\kappa} \cdot \boldsymbol{\xi}. \tag{A.13}$$

Here, one can see the contribution of the correction of the final term in (4.7).

The gyrophase-average of the second term on the right-hand side of (4.54) yields $i\omega_d \boldsymbol{\xi} \cdot \nabla F_{g0}$. One can prove that this term is of order ω_*/ω compared to the first term in (A.12) and therefore should be kept.

The gyrophase-average of the third term on the right-hand side of (4.54) is of order $(\rho^2/\lambda_\perp \lambda_\wedge)(L_B/L_p)$ and therefore can be neglected.

Inserting the quasi-neutrality condition (4.66) into the forth term on the right-hand side of (4.54), one can see that this term becomes of the same order as the $i\omega\delta G_0$ term and therefore should be kept.

The gyrophase-average of the fifth term on the right-hand side of (4.54) is of order $\lambda_\wedge/L_B$ compared to the forth term and therefore can be neglected.

The gyrophase-average of the sixth term on the right-hand side of (4.54) is of order $(\omega_*/\omega)(\lambda_\wedge/L_p)$ compared to the first term on the right-hand side of (A.12) and therefore can be neglected. Here, it is noted that $\nabla \cdot \delta\mathbf{B} \sim B/R$. This is based on the same reason as neglecting the first harmonic contribution of the eighth term in appendix A.1. Although in developing the interchange mode theories it is allowable to consider $\lambda_\wedge \sim L_p$, we do not intend to include the FLR corrections to the modes with long geodesic wavelengths $\lambda_\wedge$. The diamagnetic frequency is inversely proportional $\lambda_\wedge$ and therefore is small in this case.

The gyrophase-average of the seventh term on the right-hand side of (4.54) yields

$$\left\langle \frac{1}{B}\mathbf{v}_\perp \cdot \nabla_X F_{g0} \mathbf{e}_b \cdot \delta\mathbf{B}(\mathbf{x}) \right\rangle = -\frac{v_\perp^2}{\Omega}\mathbf{e}_1 \cdot \nabla_X F_{g0} \mathbf{e}_2 \cdot \nabla\left(\frac{1}{B}\mathbf{e}_b \cdot \delta\mathbf{B}(\mathbf{X})\right).$$

This term is of order ω_*/ω compared to the first term in (A.12) and therefore is kept. Here, an order estimate, $\delta\mathbf{B}_\parallel \sim B\nabla \cdot \boldsymbol{\xi}_\perp \sim (B/L_B)\boldsymbol{\xi}_\wedge$, is employed. This applies to the modes with frequencies much lower than the compressional Alfvén mode frequency. If the mode frequencies are of the same order as the compressional Alfvén mode frequency, one has $\omega \gg \omega_{*i}$ and this term becomes negligible.

The eighth term on the right-hand side of (4.54) is of order $(\omega_*/\omega)(\lambda_\wedge/L_p)$ compared to the forth term on the right-hand side of (4.54) and therefore can be neglected. This is based on the same reason as neglecting the sixth term.

The ninth term on the right-hand side of (4.54) is odd in $v_\parallel$. Therefore, in the following ordering estimate the $v_\parallel$-odd enhancement in the discussion of (A.11) needs to be taken into account. The $i\omega\delta\mathbf{A}$ part of the ninth term is of order $\lambda_\wedge/a$ compared to the first term on the right-hand side of (A.12). The $\nabla\delta_x\varphi$ part is of order $(\omega_*/\omega)^2(\lambda_\wedge^2/\lambda_\perp L_B)$ compared to the forth term on the right-hand side of (4.54), taking into consideration that $[(\mathbf{e}_2 \cdot \nabla)(\mathbf{e}_1 \cdot \nabla) - (\mathbf{e}_1 \cdot \nabla)(\mathbf{e}_2 \cdot \nabla)]\delta\varphi \sim (1/\lambda_\perp L_B)\delta\varphi$. Therefore, the ninth term can be neglected.

Finally, the last two terms on the right-hand side of (4.54) are of order $(\omega_*/\omega)(\lambda_\wedge/\lambda_\perp)(L_p/L_B)$ compared to the first term (reduced in (A.12)) on the left-hand side of (4.54), with the $v_\parallel$-odd enhancement in the discussion of (A.11) taken into account. Therefore, this term needs to be retained. The gyrophase average of the final term yields

$$\left\langle -v_{\parallel}\frac{1}{B}\delta\mathbf{B}(\mathbf{x})\cdot\nabla_X F_{g0} + v_{\parallel}\mathbf{e}_b(\mathbf{x})\cdot\nabla_X\left(\boldsymbol{\xi}(\mathbf{X})\cdot\nabla F_{g0}\right)\right\rangle$$
$$= \left\langle v_{\parallel}\mathbf{e}_b\cdot\left(\nabla_x\boldsymbol{\xi}(\mathbf{x})\cdot\nabla F_{g0} - \nabla_X\boldsymbol{\xi}(\mathbf{X})\cdot\nabla F_{g0}\right)\right\rangle_\alpha$$
$$= \frac{v_{\parallel}v_{\perp}^2}{\Omega^2}(\mathbf{e}_1\mathbf{e}_1 + \mathbf{e}_2\mathbf{e}_2) : \nabla\nabla\left(\mathbf{e}_b\cdot\nabla\boldsymbol{\xi}\cdot\nabla F_{g0}\right)$$
$$- \frac{v_{\parallel}v_{\perp}^2}{\Omega^2}\left[\left(\mathbf{e}_1\cdot\nabla\mathbf{e}_b\right)\cdot\nabla\left(\mathbf{e}_1\cdot\nabla\boldsymbol{\xi}\cdot\nabla F_{g0}\right) + \left(\mathbf{e}_2\cdot\nabla\mathbf{e}_b\right)\cdot\nabla\left(\mathbf{e}_2\cdot\nabla\boldsymbol{\xi}\cdot\nabla F_{g0}\right)\right].$$

Collecting the results in this appendix, one obtains the gyrophase-averaged gyrokinetic equation in (4.61).

A.3 The gyrokinetic vorticity equation

In this appendix we derive the kinetic vorticity equation (4.65) from the two current projection equations, (4.63) and (4.64), using $\nabla\cdot\delta\mathbf{j} = 0$.

The contribution of the left-hand sides of (4.63) and (4.64) to the vorticity equation is reduced as follows

$$\nabla\cdot\left(\mathbf{e}_1\mathbf{e}_1\cdot\nabla\times\delta\mathbf{B} + \mathbf{e}_2\mathbf{e}_2\cdot\nabla\times\delta\mathbf{B}\right) = -\mathbf{B}\cdot\nabla\left(\frac{\mathbf{B}\cdot\delta\mathbf{j}}{B^2}\right), \qquad \text{(A.14)}$$

where the equation $\nabla\cdot\delta\mathbf{j} = 0$ has been used.

The contribution of the first two terms on the right-hand sides of (4.63) and (4.64) to the vorticity equation is reduced as follows

$$\nabla\cdot\left[\left(-\frac{gP'}{B^2} - g'\right)\left(\mathbf{e}_1\mathbf{e}_1\cdot\delta\mathbf{B} + \mathbf{e}_2\mathbf{e}_2\cdot\delta\mathbf{B}\right)\right] = \delta\mathbf{B}\cdot\nabla\sigma - \nabla\cdot\left(\mathbf{J}_{\parallel}\frac{\mathbf{B}\cdot\delta\mathbf{B}}{B^2}\right), \qquad \text{(A.15)}$$

where we have used the relation $\sigma = \mathbf{B}\cdot\mathbf{J}/B^2 = -gP'/B^2 - g'$.

The contribution of the third term on the right-hand side of (4.64) to the vorticity equation is reduced as follows

$$-\nabla\cdot\left(\mathbf{e}_2\frac{P'|\nabla\psi|}{B^2}\mathbf{e}_b\cdot\delta\mathbf{B}\right) = -\mathbf{J}\cdot\nabla\left(\frac{\mathbf{B}\cdot\delta\mathbf{B}}{B^2}\right) + \nabla\cdot\left(\mathbf{J}_{\parallel}\frac{\mathbf{B}\cdot\delta\mathbf{B}}{B^2}\right). \qquad \text{(A.16)}$$

Here, the final term cancels with the final term in (A.15).

The contribution of the third and forth terms, respectively, on the right-hand sides of (4.63) and (4.64) to the vorticity equation is reduced as follows

$$\nabla\cdot\left[\mathbf{e}_1\frac{1}{B}\mathbf{e}_2\cdot\nabla\left(P'|\nabla\psi|\mathbf{e}_1\cdot\boldsymbol{\xi}\right) - \mathbf{e}_2\frac{1}{B}\mathbf{e}_1\cdot\nabla\left(P'|\nabla\psi|\mathbf{e}_1\cdot\boldsymbol{\xi}\right)\right]$$
$$= -\nabla\times\frac{\mathbf{B}}{B^2}\cdot\nabla(\boldsymbol{\xi}\cdot\nabla P). \qquad \text{(A.17)}$$

The contribution of the forth and fifth terms, respectively, on the right-hand sides of (4.63) and (4.64) to the vorticity equation is reduced as follows

$$\sum_{\mathrm{i,e}} \frac{1}{2} m_\rho \int \mathrm{d}^3 v v_\perp^2 \nabla \cdot \left[\frac{1}{B} (-\mathbf{e}_1\mathbf{e}_2 + \mathbf{e}_2\mathbf{e}_1) \cdot \nabla \delta G_0(\mathbf{x}) \right]$$

$$= \nabla \times \frac{\mathbf{B}}{B^2} \cdot \nabla \left(\sum_{\mathrm{i,e}} \frac{m_\rho}{2} \int \mathrm{d}^3 v v_\perp^2 \delta G_0(\mathbf{x}) \right). \tag{A.18}$$

The contribution of the fifth and sixth terms, respectively, on the right-hand sides of (4.63) and (4.64) to the vorticity equation is reduced as follows

$$\nabla \cdot \left(\mathbf{e}_1 \frac{\omega^2}{B} \rho_m \mathbf{e}_2 \cdot \boldsymbol{\xi} - \mathbf{e}_2 \frac{\omega^2}{B} \rho_m \mathbf{e}_1 \cdot \boldsymbol{\xi} \right) = -\nabla \cdot \left(\rho_m \omega^2 \frac{\mathbf{B} \times \boldsymbol{\xi}}{B^2} \right). \tag{A.19}$$

The contribution of the sixth and seventh terms, respectively, on the right-hand sides of (4.63) and (4.64) to the vorticity equation is reduced as follows

$$-\mathrm{i}\omega \nabla \cdot \left(\mathbf{e}_1 \frac{|\nabla\psi|}{B\Omega_\mathrm{i}} P_\mathrm{i}' \mathbf{e}_1 \cdot \nabla(\mathbf{e}_1 \cdot \boldsymbol{\xi}) + \mathbf{e}_2 \frac{|\nabla\psi|}{B\Omega_\mathrm{i}} P_\mathrm{i}' \mathbf{e}_2 \cdot \nabla(\mathbf{e}_1 \cdot \boldsymbol{\xi}) \right)$$

$$= -\mathrm{i}\omega \frac{|\nabla\psi|}{B\Omega_\mathrm{i}} P_\mathrm{i}' \nabla_\perp^2 (\mathbf{e}_1 \cdot \boldsymbol{\xi}). \tag{A.20}$$

The contribution of the seventh and eighth terms, respectively, on the right-hand sides of (4.63) and (4.64) to the vorticity equation is negligible for electromagnetic modes.

The contribution of the eighth and ninth terms, respectively, on the right-hand sides of (4.63) and (4.64) to the vorticity equation is reduced as follows

$$-\mathrm{i}\frac{3\omega}{2} \nabla \cdot \left[\mathbf{e}_1 \frac{n_0 T_\mathrm{i} e_\mathrm{i}}{m_\rho \Omega_\mathrm{i}^2} B^{-3/2} (\mathbf{e}_1\mathbf{e}_1 + \mathbf{e}_2\mathbf{e}_2) : \nabla\nabla\left(B^{3/2} \mathbf{e}_1 \cdot \boldsymbol{\xi} \right) \right.$$

$$\left. + \mathbf{e}_2 \frac{n_0 T_\mathrm{i} e_\mathrm{i}}{m_\rho \Omega_\mathrm{i}^2} B^{-3/2} (\mathbf{e}_1\mathbf{e}_1 + \mathbf{e}_2\mathbf{e}_2) : \nabla\nabla\left(B^{3/2} \mathbf{e}_2 \cdot \boldsymbol{\xi} \right) \right]$$

$$= -\mathrm{i}\omega \frac{3}{2B\Omega_\mathrm{i}} |\nabla\psi| P_\mathrm{i}' \mathbf{e}_1\mathbf{e}_1 : \nabla\nabla(\mathbf{e}_1 \cdot \boldsymbol{\xi})$$

$$+ \mathrm{i}\omega \frac{3 n_0 T_\mathrm{i}}{B\Omega_\mathrm{i}} \mathbf{e}_2 \cdot (\boldsymbol{\kappa} + \nabla \ln B - \mathbf{e}_1 \cdot \nabla \mathbf{e}_1) \mathbf{e}_1\mathbf{e}_1 : \nabla\nabla(\mathbf{e}_2 \cdot \boldsymbol{\xi})$$

$$- \frac{3\omega P_\mathrm{i}}{2B\Omega_\mathrm{i}} \left[(\mathbf{e}_2 \cdot \nabla \mathbf{e}_1) \cdot \mathbf{e}_2 \right] \mathbf{e}_1 \cdot \nabla(\mathbf{e}_2 \cdot \boldsymbol{\xi}). \tag{A.21}$$

Here, we have noted $\nabla \cdot \boldsymbol{\xi} = -2\boldsymbol{\kappa} \cdot \boldsymbol{\xi}$. The first and third terms are of order ω_*/ω and the second term is of order $(\omega_*/\omega)(\lambda_\wedge/\lambda_\perp)(a/R)$, compared to the inertia term.

The contribution of the ninth and tenth terms, respectively, on the right-hand sides of (4.63) and (4.64) to the vorticity equation is analyzed as follows. The terms related to the diamagnetic drift are reduced as follows:

$$-\mathrm{i}\omega\nabla\cdot\left[\mathbf{e}_1\frac{2n_0Te_\mathrm{i}}{m_\rho}\frac{|\nabla\psi|}{4\Omega_\mathrm{i}^2}\frac{\partial\ln n_0}{\partial\psi}\left(1+\eta_\mathrm{i}\right)\mathbf{e}_1\cdot\nabla\left(\mathbf{e}_1\cdot\boldsymbol{\xi}\right)\right]$$
$$=-\mathrm{i}\omega\frac{n_0T_\mathrm{i}}{2B\Omega_\mathrm{i}}\frac{|\nabla\psi|\,\partial\ln n_0}{\partial\psi}\left(1+\eta_\mathrm{i}\right)\mathbf{e}_1\mathbf{e}_1:\nabla\nabla\left(\mathbf{e}_1\cdot\boldsymbol{\xi}\right),\tag{A.22}$$

$$\mathrm{i}\omega\nabla\cdot\left[\mathbf{e}_1\frac{2n_0Te_\mathrm{i}}{m_\rho}\frac{|\nabla\psi|}{4\Omega_\mathrm{i}^2}\frac{\partial\ln n_0}{\partial\psi}\left(1+\eta_\mathrm{i}\right)\mathbf{e}_2\cdot\nabla\left(\mathbf{e}_2\cdot\boldsymbol{\xi}\right)\right]$$
$$=\mathrm{i}\omega\frac{n_0T_\mathrm{i}}{2B\Omega_\mathrm{i}}\frac{|\nabla\psi|\,\partial\ln n_0}{\partial\psi}\left(1+\eta_\mathrm{i}\right)\mathbf{e}_1\mathbf{e}_2:\nabla\nabla\left(\mathbf{e}_2\cdot\boldsymbol{\xi}\right),\tag{A.23}$$

$$-\,\mathrm{i}\omega\nabla\cdot\left[\mathbf{e}_2\frac{2n_0Te_\mathrm{i}}{m_\rho}\frac{5|\nabla\psi|}{4\Omega_\mathrm{i}^2}\frac{\partial\ln n_0}{\partial\psi}\left(1+\eta_\mathrm{i}\right)\mathbf{e}_1\cdot\nabla\left(\mathbf{e}_2\cdot\boldsymbol{\xi}\right)\right]$$
$$=-\mathrm{i}\omega\frac{5n_0T_\mathrm{i}|\nabla\psi|}{2B\Omega_\mathrm{i}}\frac{\partial\ln n_0}{\partial\psi}\left(1+\eta_\mathrm{i}\right)\mathbf{e}_1\mathbf{e}_2:\nabla\nabla\left(\mathbf{e}_2\cdot\boldsymbol{\xi}\right).\tag{A.24}$$

Note that the first term on the right-hand side of (A.21) cancels with (A.22)–(A.24), noting that $\nabla\cdot\boldsymbol{\xi}=0$ at the leading order. We also note that the combination of the terms

$$\nabla\cdot\left\{\frac{2n_0Te_\mathrm{i}}{m_\rho}\frac{7}{4\Omega_\mathrm{i}^2}\left[\mathbf{e}_1\left(\mathbf{e}_2\cdot\nabla\mathbf{e}_2\right)\cdot\nabla\mathbf{e}_1\cdot\boldsymbol{\xi}+\mathbf{e}_2\left(\mathbf{e}_2\cdot\nabla\mathbf{e}_2\right)\cdot\nabla\mathbf{e}_2\cdot\boldsymbol{\xi}\right]\right\}\tag{A.25}$$

is of order a/R smaller than the diamagnetic drift term in (A.20), noting that $\nabla\cdot\boldsymbol{\xi}=-2\boldsymbol{\kappa}\cdot\boldsymbol{\xi}$. The two terms containing the second order derivatives on $\mathbf{e}_2\cdot\boldsymbol{\xi}$ are given as follows

$$-\mathrm{i}\omega\nabla\cdot\left[\mathbf{e}_1\frac{2n_0Te_\mathrm{i}}{m_\rho}\mathbf{e}_1\cdot\nabla\mathbf{e}_2\cdot\boldsymbol{\xi}\frac{1}{8\Omega_\mathrm{i}^2}\mathbf{e}_2\cdot\left(8\nabla\ln B+\frac{1}{2}\boldsymbol{\kappa}\right)\right]$$
$$=-\mathrm{i}\omega\frac{n_0T_\mathrm{i}}{4B\Omega_\mathrm{i}}\mathbf{e}_2\cdot\left(8\nabla\ln B+\frac{1}{2}\boldsymbol{\kappa}\right)\mathbf{e}_1\mathbf{e}_1:\nabla\nabla\left(\mathbf{e}_2\cdot\boldsymbol{\xi}\right),\tag{A.26}$$

$$-\mathrm{i}\omega\nabla\cdot\left[\mathbf{e}_1\frac{2n_0T_\mathrm{i}e_\mathrm{i}}{m_\rho}\frac{1}{4\Omega_\mathrm{i}^2}\left(\mathbf{e}_1\cdot\nabla\mathbf{e}_2\right)\cdot\nabla\mathbf{e}_2\cdot\boldsymbol{\xi}\right]$$
$$=-\mathrm{i}\omega\frac{n_0T_\mathrm{i}}{2B\Omega_\mathrm{i}}\mathbf{e}_1\left(\mathbf{e}_1\cdot\nabla\mathbf{e}_2\right):\nabla\nabla\left(\mathbf{e}_2\cdot\boldsymbol{\xi}\right),\tag{A.27}$$

which is of order $(\omega_*/\omega)(\lambda_\wedge/\lambda_\perp)(a/R)$, compared to the inertia term. Here, we have used $\nabla \cdot \boldsymbol{\xi} = -2\boldsymbol{\kappa} \cdot \boldsymbol{\xi}$.

Note that $\mathbf{e}_1 \cdot \delta\mathbf{B}/B \sim (\lambda_\perp/Rq)\mathbf{e}_1 \cdot \boldsymbol{\xi}$ and $\nabla \cdot \delta\mathbf{B} = 0$. The contribution of the tenth and twelfth terms, respectively, on the right-hand sides of (4.63) and (4.64) to the vorticity equation can be neglected.

The contribution of thirteenth term on the right-hand side of (4.64) is negligible, noting that $\delta\mathbf{B}_\|/B \sim \beta\boldsymbol{\xi} \cdot \nabla P$.

The contribution of the eleventh and fourteenth terms, respectively, on the right-hand sides of (4.63) and (4.64) to the vorticity equation is reduced as follows

$$\nabla \cdot \left\{ \frac{n_0 e_i}{2\Omega_i^3}\left(\frac{T}{m_{\rho i}}\right)^2 |\nabla\psi| \frac{\partial \ln n_0}{\partial\psi}(1 + 2\eta_i)\left[\mathbf{e}_1(\mathbf{e}_1\mathbf{e}_1\mathbf{e}_2 + \mathbf{e}_2\mathbf{e}_2\mathbf{e}_2) - \mathbf{e}_2(\mathbf{e}_1\mathbf{e}_1\mathbf{e}_1 + \mathbf{e}_1\mathbf{e}_2\mathbf{e}_2)\right] \vdots \nabla\nabla\nabla(\boldsymbol{\xi} \cdot \mathbf{e}_1)\right\}$$

$$= \nabla \cdot \left[\frac{n_0 e_i}{2\Omega_i^3}\left(\frac{T}{m_{\rho i}}\right)^2 |\nabla\psi| \frac{\partial \ln n_0}{\partial\psi}(1 + 2\eta_i)(\mathbf{e}_1\mathbf{e}_1\mathbf{e}_1\mathbf{e}_2 - \mathbf{e}_2\mathbf{e}_1\mathbf{e}_1\mathbf{e}_1)\right] \vdots \nabla\nabla\nabla(\boldsymbol{\xi} \cdot \mathbf{e}_1)$$

$$= \frac{n_0 e_i}{2}\left(\frac{T}{m_{\rho i}}\right)^2 \frac{\partial \ln n_0}{\partial\psi}(1 + 2\eta_i)\left[\frac{|\nabla\psi|}{\Omega_i^3}\mathbf{e}_1\mathbf{e}_1(\mathbf{e}_1 \cdot \nabla\mathbf{e}_2) - \nabla \cdot \left(\mathbf{e}_2\frac{|\nabla\psi|}{\Omega_i^3}\right)\mathbf{e}_1\mathbf{e}_1\mathbf{e}_1\right] \vdots \nabla\nabla\nabla(\boldsymbol{\xi} \cdot \mathbf{e}_1) \tag{A.28}$$

This term is of order $(\lambda_\wedge/\lambda_\perp)(a/R)$, compared to the inertia term.

The contribution of the twelfth and fifteen terms, respectively, on the right-hand sides of (4.63) and (4.64) to the vorticity equation is reduced as follows

$$\nabla \cdot \left\{ \frac{n_0 e_i}{\Omega_i^3}\left(\frac{T}{m_{\rho i}}\right)^2 \frac{\partial \ln n_0}{\partial\psi}(1 + 2\eta_i)(\mathbf{e}_1 \cdot \nabla\,|\nabla\psi|)\left[\mathbf{e}_1\mathbf{e}_1\mathbf{e}_2 - \frac{1}{2}\mathbf{e}_2(3\mathbf{e}_1\mathbf{e}_1 + \mathbf{e}_2\mathbf{e}_2)\right] : \nabla\nabla(\boldsymbol{\xi} \cdot \mathbf{e}_1)\right\}$$

$$= -\frac{n_0 e_i}{2\Omega_i^3}\left(\frac{T}{m_{\rho i}}\right)^2 \frac{\partial \ln n_0}{\partial\psi}(1 + 2\eta_i)(\mathbf{e}_1 \cdot \nabla|\nabla\psi|)\mathbf{e}_1\mathbf{e}_1\mathbf{e}_2 \vdots \nabla\nabla\nabla(\boldsymbol{\xi} \cdot \mathbf{e}_1). \tag{A.29}$$

Here, we have noted that $\mathbf{e}_1 \cdot (\mathbf{e}_2 \cdot \nabla\mathbf{e}_1) = 0$.

The contribution of the thirteenth and sixteen terms, respectively, on the right-hand sides of (4.63) and (4.64) to the vorticity equation is negligible in our ordering scheme.

The contribution of the final terms on the right-hand sides of (4.63) and (4.64) to the vorticity equation is reduced as follows

$$\nabla\cdot\left\{\frac{n_0 e_i}{\Omega_i^3}\left(\frac{T}{m_{pi}}\right)^2|\nabla\psi|^2\left[\left(\frac{\partial\ln n_0}{\partial\psi}\right)^2\left(1+\frac{15}{2}\eta_i^2+4\eta_i\right)+\frac{\partial^2\ln n_0}{\partial\psi^2}\left(1+2\eta_i\right)\right.\right.$$

$$\left.\left.+\frac{\partial\ln n_0}{\partial\psi}\left(2\frac{\partial\eta_i}{\partial\psi}-\eta_i\frac{7}{2}\frac{\partial\ln T}{\partial\psi}\right)\right]\left[\mathbf{e}_1\mathbf{e}_1\mathbf{e}_2-\frac{1}{2}\mathbf{e}_2(\mathbf{e}_1\mathbf{e}_1+\mathbf{e}_2\mathbf{e}_2)\right]:\nabla\nabla(\boldsymbol{\xi}\cdot\mathbf{e}_1)\right\}$$

$$=\frac{n_0 e_i}{2\Omega_i^3}\left(\frac{T}{m_{pi}}\right)^2|\nabla\psi|^2\left[\left(\frac{\partial\ln n_0}{\partial\psi}\right)^2\left(1+\frac{15}{2}\eta_i^2+4\eta_i\right)+\frac{\partial^2\ln n_0}{\partial\psi^2}\left(1+2\eta_i\right)\right.$$

$$\left.+\frac{\partial\ln n_0}{\partial\psi}\left(2\frac{\partial\eta_i}{\partial\psi}-\eta_i\frac{7}{2}\frac{\partial\ln T}{\partial\psi}\right)\right]\mathbf{e}_1\mathbf{e}_1\mathbf{e}_2\vdots\nabla\nabla\nabla(\boldsymbol{\xi}\cdot\mathbf{e}_1). \qquad \text{(A.30)}$$

Collecting the results in this appendix, one obtains the kinetic vorticity equation in (4.65).

◎

编辑手记

本书是一部版权引进的英文物理学专著,中文书名可译为《先进的托卡马克稳定性理论》. 本书的作者郑林锦博士是一位华人,他是可控热核聚变等离子体领域的理论物理学家. 他在中国科学技术大学获得硕士学位,在中国科学院北京物理研究所获得博士学位,目前在德克萨斯大学奥斯汀分校聚变研究所工作. 他在《物理评论快报》《物理快报》《核聚变》《等离子体物理学》和一些大型会议上发表了一百多篇科学论文. 他的研究方向包括理想、电阻磁流体力学的托卡马克平衡及其稳定性理论. 他与同事的主要贡献包括重新计算回旋动力学理论,发展了边界局域模的物理解释,发现了二阶 TAE 本征模以及电流交换撕裂模,他还编写开发了 AEGIS 和 AEGIS-K 两个程序.

正如本书作者在前言中所指出:

> 爱因斯坦(Albert Einstein)最著名的名言之一是:世界上最不可理解的事情是,世界是可理解的. 确实,自然是受规律支配的. 由于存在自然规律,我们的科学努力才变得有意义. 人类的科学发现也可以被认为是遵循这样一条规律:要吃面包就必须汗流满面(《创世纪》3:19). 一个新发现为人类带来的“面包”越多,人类为发现它付出的“汗水”就越多. 人类面临的最严重的挑战是食物的短缺和能源的减少. 如果能够实现自然光合作用,

粮食短缺的问题就会得到解决.如果可控核聚变能以净能源产量实现,能源问题也将得到解决.不幸的是,与那些不那么重要或可能对自然环境有害的新发现相比,实现这些目标的难度更大.

可控核聚变的托卡马克等离子体理论最困难的地方在于,它把物理学哲学背后的简单和美丽弄丢了.由于许多长相关长度的带电粒子问题的复杂性与环面几何相关的复杂性的存在,托卡马克物理学在本质上是困难的.但是得益于科学家们在该领域的理论和实验方面所付出的数十年的努力,现在的托卡马克物理学变得比以往任何时候都更容易理解.本书概述了托卡马克物理学的发展状况,重点介绍了环形几何学中的稳定性理论.

这本书的目的是介绍先进的托卡马克稳定性理论.阅读本书的前提是,读者应掌握等离子体物理学的基本背景知识.我们从求导格拉德—沙弗拉诺夫方程和各种环形通量坐标的构造开始讲起.本文用解析托卡马克平衡理论证明了沙弗拉诺夫位移,以及环形箍力是如何通过托卡马克的垂直磁场被平衡的.在论述理想磁流体力学(MHD)稳定性理论时,这本书从先进且最基本的主题开始:理想 MHD 方程和 MHD 模谱的结构.其次是交换模的环形理论、气球模、剥离和自由边界气球模、TAE 本征模和动力驱动模(如动力气球模和高能粒子模).对于阻性 MHD 稳定理论,我们首先介绍了格拉瑟、格林和约翰逊的阻性 MHD 奇异层理论,之后介绍了电阻气球模和撕裂模理论,并且介绍了回旋动力学理论.我们还在回旋动力学框架下讨论了静电模式和电磁模式:首先,给出了固定和自由边界离子温度梯度模理论;然后,研究了有限拉莫尔半径对环形交换模式的影响;接着讨论了动力学气球理论和动力学 TAE 理论.

除了这些先进的理论,这本书还讨论了各种通过实验观察到的现象的直观物理图像.这些图像对理解托卡马克放电特别重要.首先,我们介绍了各种托卡马克约束模式的分类原理图,如低约束模式、高约束模式(L-模式和 h -模式)和改进之后的能量约束模式(i -模式).然后,我们讨论了各种核心和边缘稳定性、运输现象的物理解释,如输运垒、非局域输运、边缘局域模、团迹输运和边缘谐波振荡.

随着 ITER 的即将完成,以及托卡马克诊断技术和放电控制方法的迅速发展,等离子体物理理论工作者面临着巨大的挑战.这一领域仍有许多关键问题有待解决.我希望这本书有助于推动这些问题的进一步解决.

本书是一本内容较新的物理学专著.成书于 2014 年,具体内容如下:

术业有专攻,尽管要求编辑要是个杂家,但也要有一个边界,在中国经济早年间很发达但现在却又很不发达的北方,各行各业都不会给出理想的薪酬的状态下,要想各学科门类都配上相应层次、相应专业的编辑是不现实的,以笔者可怜的知识储备很快就发现本书的内容早已超出了笔者知识的边界,所以这种情况下要想硬说点什么专业的介绍一定会贻笑于方家,因此笔者不准备冒这个险.但鉴于作者是从事核聚变方面的专家,所以谈点大众都了解的原子核物理的东西似乎是可行的.正巧笔者手头有一篇60多年前,苏联的Д. В. 斯考别里金(Скобельцы)院士写的《最近二十年来人工放射性的发现及其在物理发展中的作用》似乎可以一读,故附于后.

今年——1954——是物理史上极有意义的一年,即人工放射性发现的第二十周年纪念.

在20世纪30年代,不愧为光荣不朽的玛丽亚和比埃尔·居里的后嗣——伊伦和弗雷德里克·约里奥·居里开始了自己著名的钋辐射研究.钋在谱系排列中是跟随着镭衰变而变化的最后放射性产物.在他们整个卓越工作的过程中,于1934年获得了开创原子核物理新纪元的发现,即人工放射性的发现.

这些工作及其所获得的发现属于物理学史上,甚至也可说是科学史上空前未有的成就.

在2~3年内,在世界科学中几乎同时(当然,并非完全彼此独立地)发生着头等重要的事件.不仅对物理学,而且对整个科学命运起决定性影响的发现是一个紧接着一个.

在为各方面研究而开辟的广阔道路上,集中着实验家们很大的力量,他们的工作获得新的成绩,并为科学宝库提供了新的卓越的成果.

这时期的一系列发现好像链式反应似的发展着.在科学上这一连串过程所引起的明亮光辉,在以后的日子里逐渐地失去色彩.在20世纪30年代末,特别是在战后,向核子物理科学进军的战线扩大到空前未有的规模.有关核子物理的知识已起了根本的变化.现在已达到如此高度,以致眼前展开了一片茫无际涯的视野,并且显示出不久以前还仿佛是空想的广阔前景.

如果要估量一下我所提到的,使得这个过程蓬勃推进和使得核子研究战线扩大的发现的作用,那么,不容争辩,人工放射性发现的影响是最有力的.

让我们回到发现的本身来谈谈吧.

在1932~1934年几乎同时一个跟着一个地发现了新的微粒——正

子和中子以及新的基本现象——人工放射性现象. 当然,这些"双生子"的同时问世实在不是偶然的.

人工放射性发现的本身直接与科学战线上的活跃有关,此时在科学战线的领域内要揭破正子产生和变化的现象的实质,并且这一领域是在利用宇宙线进行"探索",发现这些新微粒在自然中存在之后才产生的.

人工放射性的发现不是偶然的,而是研究者(伊伦和弗雷德里克)多年卓越工作的成果,他们在研究正子现象以及导致发现中子的一些现象方面做出了巨大的贡献.

中子,正如现在我们所知道的一样,是核子变化最有效的"刺激物",而人工放射性是核子反应最忠实的伴侣(不可分的伴生现象).

正是由于研究铀内中子所引起的人工放射性问题①,甘(Ган)和斯特拉斯曼(Штрассман)于 1939 年有了新的发现,即发现中子所引起的铀"分裂".

这个发现(在很大程度上由于弗雷德里克和他学生的新的基本工作和思想)引致利用所谓原子能(核内能)的方法. 在开辟人类掌握自然力量的新纪元之后,这个发现也有决定命运的影响,形成了一种不仅对科学的命运,而且对文明本身的存在以及人类生存的威胁.

科学世界上辉煌的事件——"双生子"中子和正子的诞生——以及与这些微粒密切相关的人工放射性的发现(像在著名的"彼罗"神话中一样)好像被咒语化为一片黑暗. 从那时起这些"咒语"的影响就威胁着今天科学的命运.

在科学中,像在人民生活中一样,随着这些著名的发现而来到的新纪元标志着矛盾的尖锐化,针锋相对的原则的斗争,创造和破坏力量的斗争,战争与和平力量的斗争……

我们所提到的正子以及与其有关的现象起了指路明灯的作用,从而使弗雷德里克首先做出证实核子反应有可能存在的实验,这些核子反应的产物是各种不同的、新的、在自然界中尚未发现有现成的形态的放射性物质.

弗雷德里克极有成效地采用了早在研究宇宙线时就发现有显著可能性的和所得的结果明确而易喻的方法.

他们用磁场内的威尔逊雾室来进行观察. 在雾室内可以看到每一个独立微粒所遗留下来的痕迹,并且在雾室内正子的痕迹直接按其磁场所引起的偏转性质来鉴定. 在具有已知的条件时,在一定的场合下正子是人

---

① И. 居里和 П. 萨维契(Савич);Л. 梅脱涅尔(Мейтнер)和 О. 斯特拉斯曼(Штрассман).

工放射性的指示者.

当在威尔逊雾室内,以 α 粒子轰击金属铝形成人工放射性之后,由于所应用观察方法的特点,就完全确实地发现了这种指示剂.

这个发现是由两个阶段完成的.在威尔逊雾室内所发现的正子起初被解释为:在铝原子核内,受钋的 α-射线轰击,而这些原子是被 α-粒子射中时所引起反应的生成物.

这个问题已于 1933 年 6 月 19 日在《法国科学院报告》所刊载的文章中作过报道.

然而,后来很快就被注意到:在引起正子放射的因素,即钋的 α-粒子放射消除之后仍能观察到正子的放射.

由此证实了:由于氦原子核即 α-粒子穿入铝、硼、镁的原子核形成新的磷、氮、矽(P^{30},N^{13},Si^{27})的同位素的放射性核子.具有很大意义的是很快就成功地用化学方法将新得出的放射性物质与形成放射性物质的物质分离开来.

随着这些最初的人工地受活化的物质而来的是更多的其他物质.

“首先并最终地被确证了可能以外部原因所引起的某些原子核的放射性,即使在其受激的原因消除之后的测定过程中,这种放射性依然保持着.”(摘自《法国科学院报告》1934 年 1 月 15 日)

得出上述结果的作者概括了他们发现的实质.但是在所引用的文章中,只是着重指出这一卓越发现的其中一方面,就实践方面而言,也许是最重要的一方面.

然而也有另外的一方面,它们的原则性意义也是不容忽略的.

正同我们着重指出的一样,弗雷德里克所获得的人工放射性同位素在衰变时所放射出来的微粒乃是正子,也就是与一般的负电子的区别只是电荷符号不同的微粒.

因此,在发现放射性物质有可能被人工合成的同时,实质上,也就是发现了新的放射性,即正子的 β-放射性.

在纯理论方面,这意味着我们对有关放射性变化的本性的概念在领域上有极重要的扩大,以及在电子的量子理论学说发展中迈进了一大步.

从实验的观点来看,也就是发现了研究正子现象的新的可能性,因为它给予研究者新的更强有力的放射正子的源泉,这一点作者已在上述最初的评语中着重指出过.

众所周知,正是由于量子理论和相对论思想的逐步发展和结合,才得出正子的理论概念.在这个基础上建立了某种粒子的理论,特别是这类粒子也包括电子,由著名的狄拉克(Дирак)方程式写出,这一理论导致粒子与其“反粒子”的对比的思潮.粒子带有电荷时,这是正的和负的粒子(特别是正子和负电子,其间的区别只是彼此电荷符号不同).

有关正子的概念像有关“反电子”的概念一样,早已成为核子物理概念的主要内容,并且完全成为一般性概念了.然而,在弗雷德里克进行初步工作的时期,它是富有开创性的.

现在,当观察家们在宇宙辐射中发现成套的新基本粒子时,诚然,是指特殊的——不稳定的(实质上,也就是放射性的)粒子,将新粒子接受为物理概念的“不动产”已不再是如此特殊的事情了.

然而,在“认识”新基本粒子方面(这种粒子不同于我们早已知道的一般粒子),好不容易才踏上最初的一步.

早在1928年~1930年发表的狄拉克理论,其思想的逐步发展得出关于正子存在的结论.但是,当时对这结论本身有怀疑的地方,而理论的创造者最初企图寻找这样的解释:使它可以避免承认一些假想的新粒子的存在.

只是当K.安德生(Аидерсен)和勃列克特(Влеккет)以及奥基阿利尼(Оккиалини)在宇宙辐射中发现正子存在之后,正子才在物理上得到公认.

由于弗雷德里克和其他科学家的工作以及在很大程度上由于开始利用人工合成的放射性物质来产生正子,才有可能详细的研究正子的现象.

正子是负电子的“对偶”,它保证自然中发生着的以及关系着电荷的量子交换的基本过程之对称(在电荷符号方面).

受一般β-衰变的中子变为质子,同时放射出在这种情况下产生的负电子(一般β-衰变的基本作用).

相反的,质子(如果它能吸收为此所需的核子能)也可以变为中子和放射出的正子(弗雷德里克所发现的β-变化).

正电荷的放射(在放射性变化的这一后期中)按照理论相当于等量负电荷被吸收.

由于弗雷德里克的实验确定了放射正子的β-衰变的存在,也就可以预料到其他类型变化的存在(建立在理论的见解上),实际上,也就存在着与一般β-变化相反的变化,即原子核内的一个质子由外界吸收得到一个普通的负电子而变为中子.

被核子捕获的并且能够引起所述变化的电子(根据理论计算)可能是最接近原子核的旅伴,也就是那两个所谓K-电子中的一个.在原子结构中,K-电子构成包藏了它们的核的电子壳的最内层.

现象的本身(弗雷德里克所发现的正子衰变的直接类推),所谓与相应的放射性变化有关的"K-层捕获",是在1937年~1938年被阿尔瓦烈茨(Алbварец)和威廉斯(Вилbямс)发现的,并且现在在核子物理中起着极重要的作用."K-层捕获"正同正子的β-衰变一样,是关系着一个元素的原子变为另一元素的原子,即在门捷列夫(Менделеев)的元素周期系统中按顺序排列的前一元素的原子.

在最近几年来正子现象的研究增添了新的极有兴趣的观察.这就是,气体(例如氮气)以放射性钠(Na^{22})所发射的正子流并在其内部吸收正子,有可能使这微粒与电子结合.在一定已知的条件下,遇到电子时正子可能与它相结合,并与其构成量子系①,此量子系实质上与氢原子相类似,但几乎比它轻2 000倍.

原来,这两个相互作用的对偶(正子和负电子)的运动的某些量子特征在顺利的情况下保证了这个系统——人工电子气体的原子——的相对稳定性,人们称这种电子气体为"电子—正子偶"(позитроний)——德文1951年.

这种相对的稳定性对于这种人工的,只是由电子构成的原子仅在极为短促的时间内(共约亿分之一秒左右)保证其可能存在.然而,这时间由于利用现代的观察技术,足以用来发现这种人工的极不稳定的气体的形成以及研究它的特性.研究这气体(正态电子—正子偶②)的性状及其衰变,这种衰变关系着形成和发射三个电磁辐射的量子即高能量的光子,从量子现象的物理观点来看是很有兴趣的.

伊伦和弗雷德里克在1934年最初发现的新型放射性在各国引起了许多卓越的研究工作,从而导致最后发现了利用"原子"能的道路,关于这点我们已经说过了,并且也导致发现了有头等意义的新的核子现象.

在弗雷德里克的发现之后不久,人工放射性和放射性示踪及分析的方法在意大利极有成效地被费米(Ферми)和他的同事用来研究中子.

费米和他的同事发现了:当中子减到很小的(相对地说)速度,达到分子热运动的速度时(每秒钟1千米左右),中子就具有一种极愿意与原子核起反应的能力,同时又表现出与原子相结合的倾向.减缓到这样的速度时,中子就变成人工放射性极有效的刺激物.除此之外,在核子与中子形成结合的倾向中,还发现了一种固定的极鲜明的并且与该核子结构的个别特点有关的选择性.进入慢中子流的原子核从这个中子流中挑选出某

① 由于在两个微粒的电荷——正的和负的——之间吸力的作用.

② 其总转矩为1之正子—电子偶谓之正态的(Ортопозитроний);如总转矩为0之正子—电子偶谓之对态的(Парапозитроний)——译者注.

些极准确的固定速度运动着的中子(用每个该种核子所特有的速度)并渴望捕获之.

被介质的原子核所捕获的中子的能级和速度,由于选择性的即共振的吸收,依靠吸收剂的本性在一定的范围内而有所变更.例如,镉以其在"热运动速度域"内能选择吸收中子为特征,这一事实尤其是在核子力能装置中被广泛地利用.

费米的发现使得核子物理的发展向前推进了极重要的新的一步.他的发现是紧接着伊伦和弗雷德里克的奠基工作而获得的.

人工活化剂的核子同质异构体的发现可算是新的卓越成就的另一例子,这种成就的取得与人工放射性在世界规模内展开研究是直接有关的.库里阿托夫(Куриатов)、罗西诺夫(Русинов)和其他的科学家(在苏联)在1935年在放射性溴(Br^{80})中首先观察到这种现象.问题在于介稳状态的存在,在这种状态中核子可在很长的时间内存在,在该情况下平均约为4个半小时.

如果放射性溴在基本状态中形成,那么在一定的时间内(平均将近20分钟)它就起衰变,并在绝大多数的情况下放射出负电子,从而变为氪.

如果在构成Br^{80}的核子时,它保持一定盈余的能,而且是在相对应的"受激的"即介稳的状态中得出的,那么β-衰变和使它成为氪核的变化一直保持到溴原子核失去借给它的盈余的能为止,例如,放射性电磁辐射的量子即高能量的光子,失去这些能量所需的时间平均为4个半小时左右.

在天然放射性的现象中,已知在铀的谱系中唯一相似的例子,其中两个不同的状态和相对应的两个不同的衰变期是在原子量为234(原子序数91)的镤同位素中发觉到的.

同一时期推测出:存在两种不同结构的变化——同质异构体,与之相应的给它们提出不同的名称:铀X_2和铀Z.甘莫夫(Гамов)曾发表过这样的假说:这个差别可能与核子结构中存在负电荷的质子即反质子有关.

在用人工方法得出的放射性产物中发现同质异构体现象之后,这个假说就被保留下来了.

同时,科学家们很快就弄清楚:同质异构现象具有普遍的性质.它非常广泛的散布在现在所知的各种不同元素的放射性同位素的"品种表"中.

为了阐明这种现象,维泽凯尔(Вейцзеккер)提出了肯定的假说,该假说的正确性目前尚未引起疑问.现在我们已经具有理论,在这理论的基础上这种现象广泛地用来研究原子核的量子状态.

同质异构体所指的是一定的状态即介稳状态，该核子从这种状态转为正常的状态是极不“愿意”的. 这种转化之所以困难是由于某些特殊的“禁律”. 它的实现对核子过程来说是异常缓慢的，从而，受激核子相应的长时间在介稳的状态下存在着，这时间对于以放射 α-线的光子或是核子—同质异构体的原子壳的某一电子实现所谓同质异构转化是必需的.

根据维泽凯尔的学说，如果在两种状态(介稳的和正常的)的能量的差别很小而同时表征核子自转的量子数差很大，即如果在正常核子与其同质异构体自转的状态中有重要的区别时，这种“延缓作用”的辐射过程的可能性在同质异构转化过程中可以得到阐明.

核子同质异构现象的意义联系到核子结构的问题在最近才开始弄清楚. 刚才所述的看来是很难结合的要求，根据核子构成的新型学说是可以被阐明的. 其中，很可能表现出某些构成核子系统和核子力规律的特征，这些特征是由最新的研究来揭破的.

我们引证了一些与应用人工放射性密切相关的发现和观察的个别例子.

不过，我们所涉及的只是现象的有限范围罢了.

在物理中“在现代技术基础上增加”的人工放射性，其应用的前途是宽广无涯的. 与利用加速核子微粒的最新技术和核子反应器来影响原子核的新方法相结合，人工放射性的现象按核子研究各种不同的方向不断扩大着. 而在研究核子反应的广阔战线上成为这些研究的基础. 即使想一一地追述这种扩大的主要线索都是不可能的.

然而，如果企图就由于二十年观察的积累和人工放射性的利用给关于物质的科学所带来的贡献做出结论，那么这个结论可以肯定地说在解决还在 20 世纪就提出的与门捷列夫名字有关的广大问题中是过渡到新的阶段了.

门捷列夫所提出的元素周期系统，实质上也就是总结我们有关物质的认识的宏伟大纲的基础，其范围按照这些知识的积累在不断地扩大着，但是实现这大纲还是等到我们的时代.

量子学说所给予的巨大的贡献是与波尔(Бор)的名字分不开的，这学说在实现这大纲上完成了一定的重要阶段.

可是，正如我们所知道的一样，完成这阶段所达到的成就，实质上是很肤浅的，它们只是涉及物质原子的外部法衣，所谓“电子外套”. 现在问题在于揭示原子核的结构和根据这些结构的理论阐明它们的特性.

由于核子物理，特别是放射性学说的成功，就此向着这以后的阶段过渡了.

天然放射性的发现在门捷列夫的表中已经带来了极大的变化,因为许多新元素和钋、镭、氡等的发现都是与其有关的,更因为(诚然,这是最主要的)它引起了有关元素变化和有关同位素现象的概念.

然而,人工放射性的发现在这方面得到新的有莫大原则性意义的结果.

位于门捷列夫元素周期系统边缘,占据他表中第 92 格的元素是铀(原子序 Z=92).

使中子与铀(U^{238})的核子相结合起来就能用人工放射法将中子变为新的放射性同位素即 U^{239}. U^{239}的核由于力求摆脱使其强迫接受的多余中子,靠着负电子的放射将其转化为质子. 结果产生了铀(239)的 β-衰变并形成越铀元素镎(Np),它占门捷列夫元素周期系统的第 93 位,并且它又靠着负电子的放射自发的产生如下的超铀元素——钚(Pr)Z=94.(西鲍里克(Сиборик) 1940 年)

在不同时间内,像核子反应的产物一样,用各种不同的方法得出下列的放射性超铀元素——镅(Am)、锔(Cm)、锫(Bk)、锎(Cf)以及按照最近的情报来说(1954 年),还有原子序 Z=99 和 100 的元素:钘(An)和钲(Ct).

在研究放射性的排列上是开辟了新的一页.

用人工方法合成的镎(Np^{237})是新放射性谱系创始者,这一谱系包括几个元素并以铊(Tl^{209})和铅(Pb^{209})的放射性同位素为结束. 铅(Pb^{209})的同位素再衰变形成铋(209)的稳定同位素.

这系原子的原子量可用公式 $A=4n+K$ 表示,式中 n——按这谱系不同组成者而不同的整数,而 $K=1$. 在天然放射性族中间已知的只有 3 个谱系,各以 $K=0.2$ 和 3 为表征.

同时,门捷列夫表中所留下的空格都被填满了,因为(正如我们现在所知的一样)这些相应的元素只以人工放射性同位素出现.

研究者一个跟着一个的得出这些新元素如锝(Z=43)、钷(Z=61)、砹(Z=85)等.

因此,我们对门捷列夫本人所提出那种形式的元素周期系统的认识得到重要的补充.

在这方面所考虑的物质原子的多样性是一元量的. 这里原子是按照一个参变数,即周期系统中元素的原子序(或者说,也就是按照原子核内的质子数)来区分的.

在考虑到核子结构时,应研究原子的二元量的多样性,因为它的各个代表均按两个参变数来区分,例如:Z 和 A,其中 A——原子量数(最接近原子量的整数);或是Z和N,其中N——原子核内的中子数①.

有关这种概括系统的规律性问题和同位素系统化的问题在我们的时代具有特殊的重要意义.

实现各种不同的核子反应,以及它们所引起的人工放射性的过程使得现在这种“奴克利特(Нуклид)”②系统包括1 000种以上的核子结构的代表(其中800种以上是放射性的).

这些结构的品种在最近,特别是战后的几年不断的增添着.

例如,在1937年首先发现的锝元素现在已知的就有12种同位素(除了它们的同质异构体之外).

1939年M. 贝尔(Пер)在锕的自然谱系中所发现的新元素——钫(Z=87)现在已有8个同位素了.

与人工放射性变化的自发过程相连和结合,现代“原子分裂”的技术提供了获得新种类原子核的可能.有可能用人工的方法构成新的,一般来说都是不稳定的,即放射性的核子结构.

放射性的辐射,因为它关系着一个元素的原子变为另一元素的原子,故仍是扩大充实整套“奴克利特”的可能性的因素.同时也是充实我们有关核子系统结构的知识的因素.

另外,就研究这些结构的可能性来说,放射性辐射本身已起了特别重要的作用.

由于放射性变化而被原子所抛出来的β-粒子,包括具有一定速度的电子及它们所发射出来的电磁辐射的量子即一定频率的光子都是特殊的信号.它将在企图改组的原子核中所发生的变化以及原子核循序产生所经过的那些状态告知观察者.靠现代核子光谱学的完善方法来发现和解释的这些信号提供了核的量子状态的最有价值的证据.现在用这种方法已积累了大量有关核子量子特征的知识的材料,而且已经在这些材料的基础上得到了许多综合性的结论.

然而,解决概括地综合这些资料的最重大的问题还摆在我们的面前.

由于这样的综合,一方面,将会揭破核子力的真相,另一方面,将会完全弄清楚核子结构的规律性,从而解决化学元素演化的问题以及其他,这样的综合将帮助我们在门捷列夫所奠基的宏伟建筑物中进入极重要的新阶段.

---

① 同Z不同N的原子称为同位素;同A不同N和Z的原子称为同量异位素.

② “奴克利特”系近来在专门书籍中经常用来表示核子结构的术语,这种核子结构同时以Z和N这两个值来表示特征.

我们所涉及的问题(即使这些也许也是简短而且片面的)是关于人工放射性的发现对物理学主要思想的成就和发展的影响(在有关原子核学说方面),并且关于这个影响的一个主要方面,即有关原子能利用方面我们只是顺便地提一下.

然而,人工放射性在各种不同的知识和技术领域内获得极广泛的应用,这是众所周知的.

由放射性原子在其衰变时由穿透性的辐射形式而发出的信号使我们可以观察这些原子,并使其用作"曳光弹".因而,它们使我们能在化学、生物学、冶金学等学科中观察各种不同过程的变化.

放射性指示器的方法即示踪原子的方法在所有这些领域内以广泛的战线发展着并成为解决最复杂问题的强有力的工具,这些问题,例如物质如何变为有机体、光合作用、扩散作用(特别是固体的扩散作用)、检查各种不同的工艺过程,研究化学反应和冶金过程等.

以其获得大强度能源的可能性与其使携性相结合,放射性辐射本身广泛地被利用着,例如在医学技术上(用来透视材料和成品以供探伤之用)、在地质勘测中、在食品的储存中、在测量技术中等.

然而,我们在这里不得不仅限于列举的例子,而且还是偶然的,因为这些问题超出本章题目的范围,有关这些问题可以写成(而且已经写成了)成卷的书籍.

实际上,最有前途和对人类造福能做出最有价值的贡献的,是在生物学上和与生物学有关的实用知识和技术的范围内——医学和农业技术上运用人工放射性原子.

在这方面最有重要意义的是:现在已可以获得如碳、磷、碘等也就是包含在动物和植物有机体内并滋养它们的物质的那些元素的人工放射性同位素.

在研究生命世界主要过程以及为了寻找掌握影响这些过程的方法的道路时,这种放射性同位素像指示器一样可以并且已经给予了不可估量的贡献.

除此之外,例如,放射性碘作为辐射的来源广泛地用作医疗剂(特别是对治疗恶性甲状腺病有特别的功效).

然而,所有这些物质在大量使用时也可能用来毁坏有机体、破坏人和动物体的组织,它们可用作最烈性的毒药.

从一方面来看,它为人类免受疾病创造了条件,另一方面,这也是大规模毁灭性的杀人武器.

天然的和用人工放射性方法合成的物质(由于利用中子)可以用来释

出核子能,也就是所谓“核子燃料”.此乃人类良好的动力来源,但同时也是具有惊人破坏力的爆炸物.

现在科学的进步所造成的这些矛盾不能再继续被容忍了.

它们的消灭有赖于人类的理性和现在团结在为和平而斗争的旗帜下的世界各国善良人民的意志.

在本文开始我们就提到的那些今天正威胁着学者们的重大发现的“咒语”应该去掉.

将科学的进步用于给人类带来无穷无尽灾难的战争的可能性和威胁应该永远的消除.

国际合作和联系的道路能使人类和科学免受这种威胁,并能创造一切的条件全力地使科学只用来为人类谋福利.这种道路是能够求得的也即将会求得的.

在为达到这目的而奋斗的战士的先进的队伍中有着这些光辉的活动家和出色的学者,像伊伦和弗雷德里克,他们是能求得这种道路的保证.他们给科学开辟了最广阔的,通向知识的光辉顶点的道路.这条道路也就是我们在本文企图描述的.

(译自苏联《哲学问题》,1954 年第 4 期,杨炜坤译,琴华校)

前面的科学知识今天读来可算作是科普了,而之后的几点议论可以看作是对今天的预言和警示.

从内容上分本书还是属于等离子体物理学,曾经国家自然科学基金资助过类似于本书主题的一个课题是:太阳系等离子体中的磁场重联.主持人也是中国科学技术大学的王水教授,不知与本书作者是否有学术方面的传承与交流.

磁场重联是太阳系等离子体物理学中一个十分重要的研究课题,也是空间等离子体和实验室等离子体中普遍存在的基本物理现象.在日冕中,它可将储存在大尺度磁场结构中的磁能快速释放,导致耀斑的发生;小尺度磁场重联时日冕等离子体加热起着重要作用,并会引起等离子体的外向流动.在行星磁层中,磁场重联形成开放型的磁层磁场位形,导致太阳风动量进入磁层的有效耦合,以及沿着开放磁通量管的等离子体质量交换.在磁尾中,尾瓣之间的快速重联引起大尺度磁场位形的变化,并伴随着激烈的等离子体动力学过程,导致磁层亚暴的发生.在等离子体彗尾中,向阳面和背阳面的磁场重联,能产生彗尾的断尾事件.在托卡马克聚变装置中,磁场重联能引起等离子体约束的瓦解.在天体等离子体系统(如吸积盘)以及星际和星系际空间中形成的各种电流片界面的区域中,磁场重联过程也可能起着重要的作用.

磁场重联的概念,最早是由 Giovanelli 提出的.他认为在磁场强度为零的中性点或中性线附近会出现放电现象,并可能对太阳耀斑的发生起着重要作用.1958年,Dungey 首先引入了重联(reconnection)一词,随即将其应用于地球磁层,建立了第一个开放的磁层模型.后来,人们提出了几种稳态磁场重联模型,并理论预言在向阳面磁层顶区存在着旋转间断和等离子体高速流.这个预言被 ISEE1 和 ISEE2 两颗卫星的同步观测结果所证实.这也提供了向阳面磁层顶区发生磁场重联的间接证据.Heos2 卫星的磁场和等离子体测量结果,表明在极尖区存在着局部磁场重联.ISEE 卫星观测到的通量传输事件(FTE),也是一种瞬时的局地重联过程.

本书因其内容艰深读者面窄,所以印量小、价格高.低定价、大折扣、高印量的时代已经结束,薄利多销的做法值得商榷了.而错位发展,是“小批量、高利润”的道路,本书甚至此系列都算作一次尝试.

今天是大年初三,在万家灯火时,枯坐办公室写出这几行文字,希望能有读者读到!

刘培杰

2021 年 2 月 19 日

于哈工大

刘培杰物理工作室
已出版(即将出版)图书目录

序号	书　　名	出版时间	定　价
1	物理学中的几何方法	2017－06	88.00
2	量子力学原理.上	2016－01	38.00
3	时标动力学方程的指数型二分性与周期解	2016－04	48.00
4	重刚体绕不动点运动方程的积分法	2016－05	68.00
5	水轮机水力稳定性	2016－05	48.00
6	Lévy 噪音驱动的传染病模型的动力学行为	2016－05	48.00
7	铣加工动力学系统稳定性研究的数学方法	2016－11	28.00
8	粒子图像测速仪实用指南:第二版	2017－08	78.00
9	锥形波入射粗糙表面反散射问题理论与算法	2018－03	68.00
10	混沌动力学:分形、平铺、代换	2019－09	48.00
11	从开普勒到阿诺德——三体问题的历史	2014－05	298.00
12	数学物理大百科全书.第 1 卷	2016－01	418.00
13	数学物理大百科全书.第 2 卷	2016－01	408.00
14	数学物理大百科全书.第 3 卷	2016－01	396.00
15	数学物理大百科全书.第 4 卷	2016－01	408.00
16	数学物理大百科全书.第 5 卷	2016－01	368.00
17	量子机器学习中数据挖掘的量子计算方法	2016－01	98.00
18	量子物理的非常规方法	2016－01	118.00
19	运输过程的统一非局部理论:广义波尔兹曼物理动力学,第 2 版	2016－01	198.00
20	量子力学与经典力学之间的联系在原子、分子及电动力学系统建模中的应用	2016－01	58.00
21	动力系统与统计力学:英文	2018－09	118.00
22	表示论与动力系统:英文	2018－09	118.00
23	工程师与科学家微分方程用书:第 4 版	2019－07	58.00
24	工程师与科学家统计学:第 4 版	2019－06	58.00
25	通往天文学的途径:第 5 版	2019－05	58.00
26	量子世界中的蝴蝶:最迷人的量子分形故事	2020－06	118.00
27	走进量子力学	2020－06	118.00
28	计算物理学概论	2020－06	48.00
29	物质,空间和时间的理论:量子理论	2020－10	48.00
30	物质,空间和时间的理论:经典理论	2020－10	48.00
31	量子场理论:解释世界的神秘背景	2020－07	38.00
32	计算物理学概论	2020－06	48.00
33	行星状星云	2020－10	38.00

刘培杰物理工作室
已出版(即将出版)图书目录

序号	书　　名	出版时间	定　价
34	基本宇宙学:从亚里士多德的宇宙到大爆炸	2020—08	58.00
35	数学磁流体力学	2020—07	58.00
36	高考物理解题金典(第 2 版)	2019—05	68.00
37	高考物理压轴题全解	2017—04	48.00
38	高中物理经典问题 25 讲	2017—05	28.00
39	高中物理教学讲义	2018—01	48.00
40	1000 个国外中学物理好题	2012—04	48.00
41	数学解题中的物理方法	2011—06	28.00
42	力学在几何中的一些应用	2013—01	38.00
43	物理奥林匹克竞赛大题典——力学卷	2014—11	48.00
44	物理奥林匹克竞赛大题典——热学卷	2014—04	28.00
45	物理奥林匹克竞赛大题典——电磁学卷	2015—07	48.00
46	物理奥林匹克竞赛大题典——光学与近代物理卷	2014—06	28.00
47	电磁理论	2020—08	48.00
48	连续介质力学中的非线性问题	2020—09	78.00
49	力学若干基本问题的发展概论	2020—11	48.00
50	狭义相对论与广义相对论:时空与引力导论(英文)	2021—07	88.00
51	束流物理学和粒子加速器的实践介绍:第 2 版(英文)	2021—07	88.00
52	凝聚态物理中的拓扑和微分几何简介(英文)	2021—05	88.00
53	广义相对论:黑洞、引力波和宇宙学介绍(英文)	2021—06	68.00
54	现代分析电磁均质化(英文)	2021—06	68.00
55	为科学家提供的基本流体动力学(英文)	2021—06	88.00
56	视觉天文学:理解夜空的指南(英文)	2021—06	68.00
57	物理学中的计算方法(英文)	2021—06	68.00
58	单星的结构与演化:导论(英文)	2021—06	108.00
59	超越居里:1903 年至 1963 年物理界四位女性及其著名发现(英文)	2021—06	68.00
60	范德瓦尔斯流体热力学的进展(英文)	2021—06	68.00
61	先进的托卡马克稳定性理论(英文)	2021—06	88.00
62	经典场论导论:基本相互作用的过程(英文)	2021—07	88.00
63	光致电离量子动力学方法原理(英文)	2021—07	108.00
64	经典域论和应力:能量张量(英文)	2021—05	88.00

联系地址:哈尔滨市南岗区复华四道街 10 号　哈尔滨工业大学出版社刘培杰物理工作室
网　　址:http://lpj.hit.edu.cn/
邮　　编:150006
联系电话:0451—86281378　　13904613167
E-mail:lpj1378@163.com